Synthesis Lectures on Threatcasting

Series Editors

This series publishes short books that explore the analytical method of threatcasting, which draws on a wide range of fields to develop techniques for recognizing future threats, designing potential futures, and exposing events that could indicate the progression toward an increasingly possible threat landscape. Books in this series will cover all aspects of the threatcasting methodology for researchers and practitioners. The series will promote new dialogues across a variety of communities and keep up to date with novel developments in the area.

Brian David Johnson · Cyndi Coon

Futurecasting

Springer

Brian David Johnson
Arizona State University
Portland, OR, USA

Cyndi Coon
Laboratory5 Applied Futures Lab
Los Ojos, NM, USA

ISSN 2771-1560 ISSN 2771-1579 (electronic)
Synthesis Lectures on Threatcasting
ISBN 978-3-032-21992-3 ISBN 978-3-032-21993-0 (eBook)
https://doi.org/10.1007/978-3-032-21993-0

This Springer imprint is published by the registered company Springer Nature Switzerland AG
The registered company address is: Gewerbestrasse 11, 6330 Cham, Switzerland

Foreword

Nearly 22,000 years ago, our ancestors painted elaborate scenes on the walls of the Lascaux caves—not just to document their present, but to imagine and communicate possible futures. These ancient markings reveal something profound about human consciousness: our unique ability to prospect potential futures and share those visions with others to inform better decisions in the present.

Today, as artificial intelligence reshapes society, climate disasters intensify, and geopolitical tensions surge, we need to reclaim and strengthen this fundamental human capacity. While virtually all of us study history, vanishingly few receive training in futures thinking—yet it is the future, not the past, that we still have the power to shape.

For the last two decades, I've been helping leaders learn how to anticipate, imagine, and shape better futures. As the Futurist in Residence at Stanford University's Hasso Plattner Institute of Design ("the d.school"), teaching classes in strategic foresight and creating experiences like "The Futures Happening: Democracy Edition," I've seen firsthand how applied futures thinking can transform abstract threats into actionable insights, action and impact.

The urgency of systematic futures thinking has never been clearer. The last five years have demonstrated the devastating cost of myopic planning. From COVID-19's cascading disruptions to AI's rapid emergence, from accelerating climate disasters to deepening social polarization—our institutions have repeatedly been caught flat-footed. This isn't just a failure of preparation; it's a failure of imagination. Traditional planning methods, focused on quarterly metrics and annual goals, leave us dangerously unprepared for the complex, interconnected challenges that define our era.

The practice of futures thinking offers a robust alternative. Its modern applications trace back to the Cold War, when Herman Kahn and the RAND Corporation pioneered scenario planning for military strategy. Royal Dutch Shell later adapted these methods for corporate planning, famously navigating the 1970s oil crisis while competitors floundered. But today's challenges demand even more sophisticated approaches. Climate change, technological disruption, and social transformation create compounding feedback loops that grow more complex with each passing year.

This is where Brian David Johnson and Cyndi Coon's work at Arizona State University's Threatcasting Lab breaks new ground. Their latest book *Futurecasting* transcends traditional scenario planning by combining creative exploration with rigorous analysis of emerging threats. Unlike conventional approaches that extrapolate from current trends, *Futurecasting* helps organizations actively shape more positive futures while preparing for adverse scenarios. Their framework extends our view beyond immediate horizons, incorporates diverse perspectives to challenge ingrained assumptions, and balances imaginative "What if?" thinking with careful analysis of what's plausible.

As cultural anthropologist Grant McCracken observed, "There is really only one way to live in a world of speed, surprise, noise, and responsiveness, and that's to visit the future frequently." Johnson and Coon's work provides a systematic approach to these vital future visits. Their book outlines a clear, step-by-step methodology that makes futures thinking accessible to anyone—from middle school students to corporate executives. Through structured exercises, visual frameworks, organizational tools, and guided reflection prompts, readers learn how to map emerging threats, identify leverage points for positive change, and develop plans for concrete actions they can take today.

For leaders seeking to build resilient organizations and educators preparing the next generation of decision-makers, the methods detailed here provide an essential foundation for systematic futures thinking and integrated skill building. A combination of rigorous analysis with storytelling and worldbuilding techniques sets their work apart, helping diverse participants move beyond abstract speculation to create tangible future scenarios they can test and refine.

This book arrives at a crucial moment, as the pace of change accelerates and the consequences of poor preparation grow more severe. Put simply, we need more leaders who are not just capable of imagining new futures, but are able to build the organizational and community capacity for others to do it, too. In an era where reactive decision-making too often prevails, Johnson and Coon offer both the imperative and the roadmap we need to navigate—and shape—an increasingly uncertain world.

Lisa Kay Solomon
Futurist, Author, Educator and Innovator
Stanford d.school

Preface

How to Design the Future—BDJ

How do we design the future? This is a question that people have been asking themselves for quite a long time. Designing the future is far more different than "seeing the future," "future-proofing your organization" or even "preparing for the future." These are all inherently passive. There is an assumption baked into these words that the future is fixed and we as humans are just running toward it, helpless to take any action.

Designing the future is different. Inherent in these words is a kind of agency, the idea that we all have the ability to not only see possible futures but shape those futures as well. This is the foundational element of the Futurecasting Method. It embraces the idea that people, individuals and organizations cannot only imagine a range of possible and potential futures but that they also as individuals and organizations have the ability to shape that future.

The origin story the Futurecasting Method is firmly rooted in system engineering or more specifically preparing to solve future problems. The method was born out of necessity, to design technological systems that would take 4 to sometimes 10 years to be developed. To do this, a rigorous, transparent data based method was needed to provide not only specific actions to be taken over that time period but also indicators along that way that the future was actually going to manifest. This was used first at VISONIK, an interactive TV design agency in Copenhagen, then adapted at the Intel Corporation for the development of Smart TV platforms and technologies.

The Futurecasting Method was also adapted from a generative approach, envisioning and developing possible solution and experience to a preventative variant called Threatcasting. Threatcasting envision possible and potential threats so that they can be identified, disrupted, mitigated or received from.

Futurecasting is a method that enables multidisciplinary groups to envision and plan systematically to identify possible futures and opportunities ten years in future. Groups explore how to transform the future they desire into reality while avoiding an undesired future. Futurecasting uses inputs from social science, technical research, cultural history, economics, trends, expert interviews, and even a little science fiction. These various inputs allow the creation of potential futures (focused on the fiction of a person in a place solving a problem). Some of these futures are desirable while others are to be avoided. By placing the futures into a fictional narrative, it allows decision-makers and practitioners to imagine what needs to be done today, as well as four and eight years into the future, to empower or disrupt the targeted future scenario. The framework also illustrates what flags, or indicators, could appear in the world that indicate the progress toward the desired future.

How to Get the Most from This Book

This book is broken into two parts. Part I will give you an overview of the Futurecasting method with examples to enhance the academic foundation. At the end of these chapters, you will find exercises to enhance your understanding of the method and prepare you to apply what you learned. Part II/Chapter 8 offers a project where you can conduct a Futurecasting project on the topic of your choice. This will enable you to pull together the knowledge from Part I and apply it. Part II of the book is designed as a practitioner's handbook.

Throughout this book, you will find Interviews. These are conversations with participants, subject matter experts, facilitators, curators, writers, academics, artists, advertising agents, and other government and military officials. These interviews are provided to offer a lens into the many different critical voices that are necessary to create Futurecasting results. They each have unique experiences with the process, and their stories are shared in order to enrich the text.

Exercise

At the end of the Part I chapters, you will find Exercises. The purpose of providing exercises is to pause and work on building your Futurecasting "muscles" before proceeding to the next chapter. Some of the exercises are considered "skill building"—focused on developing a skill or concept that will be useful when Futurecasting. Other exercises provide an opportunity to experience an aspect of the Futurecasting Method in an applied space. Each time you find an exercise, plan to run through it yourself or with a group, capture notes, and reference them when you are building your own Futurecasting models. Completing the exercises in order will help build the skills needed for the next chapter and the next.

Project

There is a project in Part II/Chap. 8. This project is a Futurecasting from beginning to end for the purposes of practicing and learning the Method. The project will contain all the elements in Part I so that the analyst(s) can rehearse, experiment, and practice the Futurecasting Method in a contained environment.

Stories from the Lab

As the interviews enhanced the foundational concepts—there are snippets of stories from the Lab where diverse, interdisciplinary collections of people and organizations are brought together to explore, research and model possible futures to encourage specific enabling actions and to foster conversations and dialogue across a broad range of audiences. These stories are derived from conducting various Futurecasting activities over the years. The purpose of these stories is to highlight lessons learned and funny situations that have occurred during the application of the Futurecasting Method to a multitude of topics for various organizations over the years.

Get Started!

The best way to get to know and use Futurecasting is to conduct Futurecasting sessions. It's like fishing: the first time you go out, someone has to show you how to put the three pieces of the pole together, attach the reel, thread the line, attach the hook and worm, and how to cast. Then there is a whole conversation about what to do while the worm is in the water. You need to fish a few times to get the hang of it. Generally, each time the pieces are the same and go together the same, but the outcomes could leave you with a big fish story!

Futurecasting is a method with two high-level segments. The first segment imagines possible futures by gathering together a broad range of multidisciplinary inputs to model a range of possible and potential futures 10 years in future. The second segment then backcasts—specifying what needs to happen over the next 10 years to enable these futures to take shape. As Futurecasting is an applied futures methodology, these actions are specific to an organization, group of people, or research agenda. This book combines method explanation (Part I) and a "how to" Project section (Part II) so that you can apply it to specific problems.

Key Terms

This book will use specific terms and descriptors to outline Futurecasting. See the glossary for additional terms used in this text.

The Use of Parenthetical Plural(s)

When specific people are referenced in the book, most will be followed by an (s) symbol or parenthetical plural. This is done to show that the task can be performed by a single person or multiple people.

Analyst(s)

The analyst(s) role in Futurecasting is to lead the effort. Typically, the lead analyst will run the effort from definition, planning, curation, execution, post-analysis, and reporting. Other analysts can be brought in to support the lead analyst in some or all tasks and to validate the work of the lead. These support analyst(s) can be a benefit if the Futurecasting is larger in scope and size. Specifically in the post-analysis stage, support analysts can be brought in to augment and broaden the lead analyst's perspective.

Participant(s)

The participant(s) are brought in by the analyst(s) to take part in the Futurecasting Workshop.

Participant(s) are typically involved in the project only during the workshop. However, some participant(s) might also take part in a peer review of the final findings after the analyst(s) conduct the post-analysis.

Subject Matter Expert (SME)

A Subject Matter Expert (SME) is a person with a particular expertise, perspective, or opinion that the analyst(s) selects as a prompt for the Futurecasting Workshop. Generally, SMEs do not participate in the workshop. They can, however, participate in a peer review of the final findings after the analyst(s) conduct the post-analysis.

Futurecasting Foundation

The Futurecasting Foundation consists of the primary topic, research question, and applications area that the Futurecasting Method will cover. It also lays out how the output will be used. This is discussed in Chap. 2. Establishing a robust, researched, and focused foundation is an important skill for any research project. The better defined the foundation, the more efficient and effective the application of the Futurecasting Method will be.

Science Fiction Prototype

Science Fiction Prototypes (SFP) are plausible science-fiction futures based on the Futurecasting research that allow the participant(s) and reader to explore the ethical, cultural, policy, and security impacts on people. The specific events depicted exist only in the imagination of the authors. The locations are selected merely to dramatize the storylines and could just as easily be any city or nation confronting external futures.

These SFPs are used as a part of the Effect-Based Model (EBM), to give participant(s) a process that elicits details about the future

Effect-Based Model (EBM)

The central device used in the Futurecasting Method is the Effect-Based Model (EBM). The EBM is a qualitative model wherein participant(s) use the prompts and the Research Synthesis Workbook (RSW) data to imagine possible and potential futures. An EBM is a structured database (e.g., spreadsheet) that poses to the practitioners a series of questions and requires a list of tasks to fill in to create the qualitative futures model. Moving from the high-level research (prompts) and participant(s)' perspectives in the RSW, practitioner(s) focus on a specific future. Moving from the macro-level to the micro-level, practitioner(s) explore a person in a place experiencing the solution to the problem.

The EBM uses Experience Design Methods and Effects-Based Operations methods, tools, and approaches to guide the participants through the futurecasting and backcasting exercises.

Project Documentation

Project documentation is a collection of files that analyst(s) use throughout the Futurecasting Method. These types of documents can be text docs (e.g., MS Word, Notepad, Google Docs/Google Forms), spreadsheets (e.g., MS excel, Google Sheets), or slides (e.g., MS Powerpoint, Slides). The type of document doesn't matter so much.

The analyst(s) uses the project documentation to keep track of every step on the futurecasting process such as:

- capturing the Futurecasting Foundation;
- generating a list of possible prompts and Subject Matter Experts;
- listing out possible participant(s);
- generating the various workbooks; and
- capturing analyst(s)' notes during post-analysis.

Diligent project documentation will allow the process to go smoothly, ease communication with others, simplify tracking, and ensure there is complete data transparency at the end of the process.

Why Poetry Matters

We start with poetry because it's weird. And weird is where the future lives.

In every Futurecasting workshop, a poem kicks off the session. Always. Sometimes it's a single line; sometimes it's an ecerpt. But there is always a poem. Why? Because poetry stands in as a reminder to focus on the humans. And it's unexpected and that's the point. You walk into a session expecting to be a part of research, and you get a string of words in a poem. It disorients. It disrupts. It tilts the room just enough to open the imagination.

Poetry isn't just flair. It's a technology that gets past the corporate scripts and expert armor. When a participant hears a poem, they're nudged out of PowerPoint mode and dropped straight into a feeling. That matters. Because if you're not feeling the future, you're not really imagining it.

We use poetry to introduce discomfort, the good kind, the kind that breaks open the present. It unmoors us from the now and makes room for the strange and the speculative. When we Futurecast, we are architecting possibilities. And possibility lives in metaphor. In contradiction. In ambiguity. Poetry models that. It holds tension. It refuses resolution. It teaches us to sit inside paradox which is exactly where the future will ask us to live.

As mentioned, poetry is also a stand-in for humanity. It reminds us that all futures are human futures. Behind every data point is a body. A breath. A beating heart. Poetry recenters that truth. It brings the people back into the room. So when we start an exercise with a poem, we are choosing to begin with meaning. We do this because the work of Futurecasting is about conjuring what could come next and then building the scaffolding to get there.

Poetry matters because the future should matter. And to matter, it must be felt.

Portland, USA — Brian David Johnson
Los Ojos, USA — Cyndi Coon

Contents

Part I

Futurecasting: Method, Framework, and Process

1 Futurecasting

I choose to make the rest of my life the best of my life.

—Louise Hay

1.1 Futurecasting: An Overview

Futurecasting is a method that enables multidisciplinary groups to envision and plan systematically for a range of possible futures and opportunities. Groups explore how to transform the future they desire into reality while avoiding an undesired future. Futurecasting uses inputs from social science, technical research, cultural history, economics, trends, expert interviews, and even a little science fiction. These various inputs allow the creation of potential futures (focused on the fiction of a person in a place solving a problem). Some of these features are desirable while others are to be avoided. By placing the futures into a fictional narrative, it allows decision makers and practitioners to imagine what needs to be done today as well as years into the future to empower or disrupt the targeted future scenario. The framework also illustrates what flags, or indicators, could appear in the progression of the future.

Futurecasting is a human-centric process, and therefore the humans that participate in a Futurecasting session are critical. Regardless of age, experience, or education, all participants are considered practitioners. Futurecasting is a theoretical exercise that can be undertaken by practitioners with special domain knowledge of how to specifically identify possible futures and opportunities and what needs to happen to bring them about. In addition, participants are curated to be outliers, trained foresight professionals, and young participants for a fresh and multi-generational perspective in the groups, artists,

B. D. Johnson and C. Coon, *Futurecasting*, Synthesis Lectures on Threatcasting,
https://doi.org/10.1007/978-3-032-21993-0_1

philosophers and public relations as examples. When using Futurecasting on military problems, the mixture of participants should span academia, private industry, government, and the military.

1.2 Futurecasting: An Ontological Discussion

To introduce Futurecasting it might be helpful to have an ontological discussion. When it comes to information science, an "ontology encompasses a representation, formal naming, and definition of the categories, properties, and relations between the concepts, data, and entities that substantiate one, many, or all domains of discourse. More simply, an ontology is a way of showing the properties of a subject area and how they are related, by defining a set of concepts and categories that represent the subject." (Ontology 2021).

For our Futurecasting discussion, we are going to review what Futurecasting is. For our purposes, Futurecasting is a method, a framework, and a process. By exploring how Futurecasting has the properties of each of these, we can get a better high-level understanding of it before we dig deeper into the details.

1.2.1 Futurecasting: A Method

> The most effective way to do it, is to do it.
>
> — Amelia Earhart

Futurecasting as a method provides analyst(s) with a series of defined steps, tasks, systems, and ultimately outputs to identify a range of possible and potential futures, along with actions to be taken and indicators to be monitored. Futurecasting is an applied futures or foresight method that not only identifies a range of possible and potential futures but specifies indicator(s) and action(s) for a specific audience. Meaning that Futurecasting Projects are conducted for a specific group or organization so that they can put them to use by identifying possible and potential futures for the specific organization so that they can take action.

Futurecasting is an analytical method that explores tomorrow's futures today, in order to give organizations and communities time to detect, prepare, disrupt, mitigate, and, when needed, recover. The goal of Futurecasting is not to predict the future. It is not a crystal ball that will definitely find the exact future picture 10 years from now. Instead, it provides a range of futures that you might not have been aware of in order to do something about it.

Example

"Automated Trust" in 2034
Research Prompt

By 2034, AI-enabled autonomous vehicles have become the backbone of urban logistics, reducing traffic fatalities and emissions. However, with these advances what are the obstacles to public trust and safety due to hacking, false narratives, and algorithmic biases?

Futurecasting Process

Vision

A bustling metropolis where drones and driverless trucks operate silently. Pedestrians trust the system for the most part until, one day, an unknown actor hacks into the central logistics AI. A series of "accidental" incidents cascade across the city, paralyzing the flow of goods and services. These "accidents" erode trust in the technology, leading to public outcry for manual overrides and localized logistics.

Futurecasting Opportunities

Opportunities for a Resilient AI-Driven Urban Future

AI Innovation for Public Safety: Develop adaptive AI systems that use predictive analytics to detect anomalies before they escalate, ensuring a safer and more reliable logistics network.

Trusted Autonomy Frameworks: Build cross-sector collaboration to establish ethical and transparent AI standards, fostering public trust and international alignment on AI governance.

Economic Resilience and Localization: Invest in decentralized logistics models that use autonomous systems in tandem with local supply chain hubs, creating redundancy and economic opportunities for smaller communities.

Opportunities to Leverage Emerging Indicators

Layered Security Architectures: Encourage widespread adoption of modular AI systems, allowing decentralized decision-making to improve resilience against hacking or systemic failures.

Algorithmic Integrity Programs: Launch global initiatives to monitor and validate AI decision-making processes, ensuring that detected anomalies lead to immediate improvements and public accountability.

Counter-Narrative Campaigns: Use social media analytics and machine learning tools to identify and dismantle disinformation campaigns aimed at eroding trust in autonomous systems.

Actionable Strategies for Building the Future

Quantum-Secured AI Ecosystems: Deploy quantum encryption and blockchain technologies to safeguard data integrity in AI-driven logistics networks.

Community-Centric AI Education: Partner with community leaders and educators to demystify AI technology, showcasing its societal benefits and safeguards.

Diverse AI Workforce Development: Expand opportunities in AI engineering and logistics for underrepresented groups, ensuring a broader perspective in system design and implementation.

The Futurecasting methodology highlights a pathway to prepare for a potential erosion of public trust in transformative technology. By identifying specific obstacles and opportunities, urban planners, policymakers, and engineers can mitigate risks and enable positive action, ensuring that the promise of AI logistics doesn't collapse under the weight of preventable vulnerabilities.

Why 10 Years in Future?

The Futurecasting Method focuses on ten years in future. This is a conscious timing decision in order to reduce innate biases from participant(s) when they participate in the Futurecasting Workshop and explore possible futures. The ten-year time horizon allows for, and overcomes, plausibility concerns. For most, envisioning ten years into the future is an intellectually freeing experience, allowing participant(s) to imagine a broader range of futures beyond their current state.

Typically, the ten-year time horizon is freeing because it is past the duration of:

- political administrations;
- corporate executives' appointment;
- the life cycle of most projects; and,
- the current career or life position of the participant(s).

Using the Futurecasting methodology, the participant(s) can free themselves from the baggage of their emotional and intellectual connections to the present, allowing them to envision the Future.

The 10-year time horizon is also a helpful guide when curating the prompts that will be used to influence the participant(s). A decade is near enough that current research applies and not so far away that the prompts would be implausible.

Many organizations and businesses will plan for the future, but their planning is generally a few fiscal quarters out or possibly two to five years. These strategic plans focus on current conditions, incremental change and business metrics. The ten-year horizon can help these planners "leapfrog" their current planning activities, expanding the range of the possible and probable futures. This expansion could identify futures that had been previously missed.

Ten years is also important as this method develops many visions of the future that could significantly change our way of life. But ten years is enough time to enable or derail these futures. Participant(s) are empowered to develop solutions to ensure that negative futures don't occur and positive futures are enabled.

The Six Phases of the Futurecasting Method

The Futurecasting Method is broken into six steps or distinct phases that contain tasks and activities. These phases are meant to provide the analyst(s) structure and guidance to conduct the Futurecasting. They need to be followed closely. A phase cannot be omitted or skipped. The tasks and activities inside of each phase need to be performed before the analyst(s) can move on to the next phase.

The phases are numbered zero through five. For some, the idea of starting with a Phase 0 (zero) might seem strange. Phase 0 is full of the preparatory actions before the Futurecasting Workshop and other activities begin. In the military, Phase 0 is defined as "shaping the environment" and contains activities designed to set the conditions for success in future phases. For technology development projects, Phase 0 is the scoping and preparation part of the project. Often this is seen as the most important phase as all the following phases build upon it.

Note: We will go into greater detail in the upcoming chapters of the book.

Phase 0: Preparation and Curation (Pre-workshop)

The beginning phase of the method consists of the preparation of the project and the workshop.

This phase also includes the curation of the team, participant(s), and research prompts.

The initial action for the analyst(s) is to develop the Futurecasting Foundation, consisting of:

- the topic area to be explored;
- the specific research question; and,
- the area(s) where the findings will be applied.

Informed and guided by the foundation, the analyst(s) pulls together a team, determines who should participate in the workshop, and decide what research or inputs should be used as prompts to envision possible futures 10 years out. Finally, the materials (e.g., workbooks, presentations, support materials) are created to conduct the workshop.

Phase 1: Prompt Presentation, Research Synthesis, and Discussion (Workshop)

The next phase begins the actual Futurecasting Workshop. Analyst(s) use the prompts and materials to engage in a participatory design session with participant(s). This activity presents the prompt to the participant(s) and then leads them through a session to explore the ramifications of the prompts, capturing their discussion in workbooks for use later by both the participant(s) in the following stages of the workshop and by the analyst(s) in the post-workshop phases.

Phase 2: Futurecasting (Workshop)
Guided by the prompts and research synthesis, participant(s) engage in another participatory design session to envision possible and potential futures ten years out. Participant(s) move from the high-level macro view of the research synthesis and prompts to the micro perspective of a person in a place experiencing a problem. To do this, they follow the Science Fiction Prototyping (SFP) and Experience Design Processes to generate a qualitative Effect-Based Model (EBM).

Phase 3: Backcasting (Workshop)
Using the EBM, participant(s) begin backcasting in small groups, developing a time-phased, alternative-action definition (TAD) phase that generates specific actions that can be taken to enable or generate positive future outcomes while detecting, preparing, disrupting, mitigating, and, when needed, recovering from the futures. Additionally, participant(s) identify the indicators (flags) over the next decade that will show that a possible future is beginning to manifest and become a reality. Phases 2 and 3 can be repeated multiple times during the workshop to generate a high volume of possible futures. The research synthesis workbooks along with the Futurecasting workbooks make up the Futurecasting Method's raw data to be processed by the analyst(s).

Phase 4: Post-analysis, Synthesis, and Findings (Post-workshop)
After the conclusion of the workshop, the analyst(s) study the raw data, using multiple techniques to cluster and identify the possible and potential futures. These findings are documented and sometimes peer reviewed by the participant(s) and Subject Matter Experts (SMEs). Additional research is conducted if needed, and the technical documentation captures the futures, actions, and indicators.

Phase 5: Output (Post-workshop)
The final phase of the method translates the findings into an output. This output is determined by the Futurecasting Foundation in Phase 0. The correct output (e.g., technical report, academic paper, podcast, etc.) is determined by the person or organization that will be applying or using the Futurecasting findings.

Applying the Method
The Futurecasting method is distinct from traditional notions of futures thinking, planning, and modeling. Not only does the methodology combine both linear and creative thinking, it also requires that a diverse set of participant(s) from multiple organizations and domain expertise gather and collaborate. This diversity of participant(s) and the multidisciplinary nature of the sources it draws upon paired with multiple guided exercises to explore possible futures enables groups to envision a complex and evolving future landscape.

Stories from the Lab

When it comes to Futurecasting workshops, diversity of thought is key—and occasionally leads to unexpected debates. In one session, a discussion about AI-driven food recommendations spiraled into a passionate debate about pineapple on pizza. The chef called it a crime, the sociologist saw globalization in action, and the policy maker suggested regulating AI to preserve culinary heritage.

It was a lively reminder that even small, unexpected topics can spark big ideas—and sometimes, that's exactly what a Futurecasting project needs.

1.2.2 Futurecasting: A Framework

> Frameworks are by definition a little loose. They exist to provide structure and direction on a preferred way to do something without being too detailed or rigid. In essence, frameworks provide guidelines. They are powerful because they provide guidance while being flexible enough to adapt to changing conditions or to be customized for your company while utilizing vetted approaches.
>
> — Scott Ellis

From Method to Framework

Once an analyst(s) understands the specific steps, tasks, methods, and, ultimately, outputs of the Futurecasting Method, they can begin to use Futurecasting as a Framework. Within the confines of the methodology, the analyst(s) can use Futurecasting as a framework to explore future areas, problem sets, and timelines.

The framework is designed to provide flexibility and serve as a guideline. The framework is the outline of tasks to execute the Method. However there is some "wiggle room" for other tools or practices to be used in order to achieve the end goals of the Method.

The Futurecasting Framework (Fig. 1.1) captures the tasks of the methodology but allows the analyst(s) to make decisions to curate and hone the activity. This will depend on the Futurecasting Foundation, specifically the applications area(s).

Example Adjustments

The goal of the framework's flexibility is to allow the analyst(s) the ability to make changes and adjustments to work with the participant(s) to investigate a wider range of futures and provide detail in areas that relate to how the results will be used.

This understanding could modify the Futurecasting Foundation or the details for how the analyst(s) conducts the workshop.

Additionally, in the midst of the workshop the analyst(s) may need to adjust a specific task, prompt, workbook, or detail to generate the results and data needed for the application area(s).

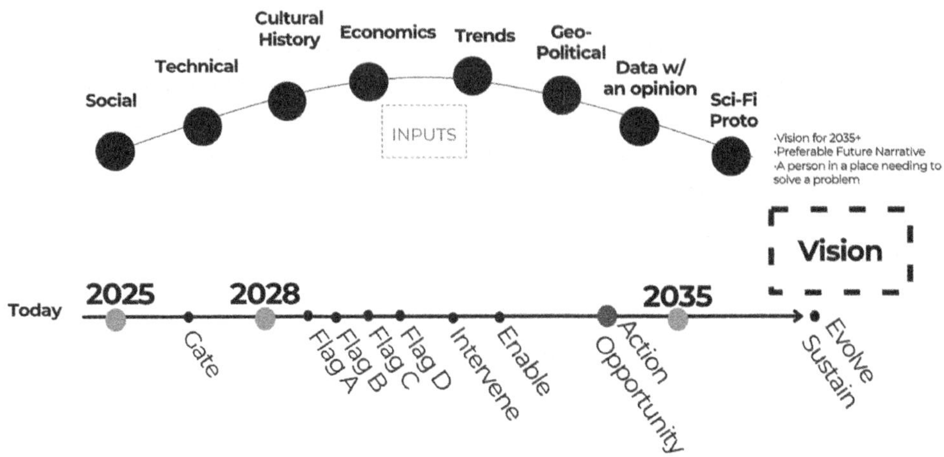

Fig. 1.1 Futurecasting framework

Example Adjustments

- Specify a future actor;
- Modifying the location;
- Limiting the prompts;
- Adding new information.

The Method remains the same, but when used as a framework, Futurecasting can adjust and flex to accommodate the specific needs of the Foundation and the necessary generation of the raw data needed for the post-analysis and findings.

1.2.3 Futurecasting: A Process

> The process is the process.
>
> —The authors

Conceptually, we have walked from a method to a framework and finally to a process. Analyst(s) can use Futurecasting as simply a process, a series of steps to be used to explore the future. It is a repeatable process. In many application areas, analyst(s) conduct multiple Futurecasting Workshops and projects, exploring a wide range of futures in different areas. Often the findings from a Futurecasting project will uncover a new area for research wherein the analyst(s) will need to perform another, more detailed Futurecasting Workshop to generate more raw data and results for their application area.

The framework is the outline of tasks while the process is the applied application of these tasks to reach a specific result. The process could and usually is different depending upon the desired outcome and the application of the findings. Part 2 of this book outlines the process for designing and executing multiple sizes and instantiations of Futurecasting. We provide a series of steps and decision points for the analyst(s) to follow when applying Futurecasting to a specific problem and organization.

Chapter Resources

Vocabulary

Futurecasting
A structured, human-centric method that enables multidisciplinary groups to envision and plan for a range of possible futures, transform desirable futures into reality, and avoid undesirable ones.

Practitioner
Any participant in a Futurecasting session, regardless of age, experience, or education, engaged in identifying possible futures and opportunities.

Ontology
A formal representation of concepts, categories, and relationships within a subject area, showing how they are related.

Method
A defined set of steps, tasks, and outputs for identifying possible futures, specifying actions and indicators for a specific audience.

Framework
A flexible set of guidelines within the method, allowing adjustments to address specific contexts while maintaining core tasks.

Process
The repeatable application of method tasks to explore the future, often in multiple workshops or projects.

Science Fiction Prototyping (SFP)
A narrative tool for imagining a person in a place solving a problem in future, used to explore ethical, cultural, and policy implications.

Effect-Based Model (EBM)
A qualitative model generated from prompts and research synthesis, focusing on specific future scenarios.

Backcasting
A method of working backward from a desired or avoided future to identify actions and indicators over time.

Indicators/Flags
Observable signs or signals that a specific future is beginning to manifest.

Time-Phased Alternative-Action Definition (TAD)
A structured plan mapping specific actions over time to achieve or avoid a future.

Phase 0–5: The six phases of the Futurecasting Method.

Resources

Primary Inputs for Futurecasting defined as:

- **Social Science Research**—Understanding societal evolution and human adaptation.
- **Technical Research**—Early-stage or emerging technology developments.
- **Cultural History**—Historical context influencing societal response to futures.
- **Economic/Geopolitical Analysis**—Macro and micro economic and political shifts.
- **Trend Analysis**—Regulatory, cultural, and market trends affecting the topic.
- **Expert Opinions (SMEs)**—Specialized insights into specific domains.

Recommendations

- Outliers with unconventional perspectives;
- Foresight professionals;
- Younger participants for multi-generational views;
- Cross-sector representatives for specific problem areas.

Useful Links

Multi-Sided Dice link Dice Shop: https://www.thediceshoponline.com/specialist-dice/multi-sided-dice-sets.html
Applied Future Lab: https://www.appliedfutureslab.com/

Threatcasting Community: Threatcasting.ai
Threatcasting Lab at ASU: https://threatcasting.asu.edu/.

Reference

Ontology (Information Science) (2021) In Wikipedia. https://en.wikipedia.org/wiki/Ontology_(information_science)

2 Phase 0

> Everything happening around me is very random. I am enjoying the phase, as the journey is far more enjoyable than the destination.
>
> —Sushant Singh Rajput

2.1 Preparation and Curation

In U.S. Department of Defense terminology, Phase 0 is a shaping phase—namely, influencing the state of affairs in peacetime before an operation. Shaping activities (occurring in this phase) help set the conditions for the successful execution of a military operation. Similarly, Phase 0 in the Futurecasting Method consists of preparation for the project as well as curation of the research that will be used and participant(s) that will be a part of the workshop. Careful preparation and curation will ensure the effectiveness and clarity of the results.

In technology product development, Phase 0 is often the first phase of a project. A Phase 0 converts the core product ideas into detailed specifications that can then be fully designed and implemented. Often during the Phase 0 project, teams collaborate with clients, and end users of the project, on the features, functions, and interfaces of a new product, generating a well-defined set of requirements. "Phase 0 of Projects provide the structure…to define the problem and evaluate potential solutions before diving into the detailed design implementation" (AppliedLogix n.d.).

The activities within this phase are:

- develop the Futurecasting Foundation;

B. D. Johnson and C. Coon, *Futurecasting*, Synthesis Lectures on Threatcasting,
https://doi.org/10.1007/978-3-032-21993-0_2

- assemble the core team, steering committee, and analyst(s);
- select and gather research prompts;
- select the participant(s); and,
- draft the Futurecasting workshop workbooks.

2.1.1 Futurecasting Foundation

> To ask the right question is harder than to answer it.
>
> —Georg Cantor

Phase 0 begins with the analyst(s) developing the Futurecasting Foundation. The foundation (see Fig. 2.1) has three main components: the topic, research question, and applications area along PHASE 0 with an understanding of how the output of the Futurecasting Method will be used. Establishing a robust, researched, and focused foundation is an important skill for any research project. The better defined the foundation, the more efficient and effective the application of the Futurecasting Method will be.

1. Topic;
2. Research question;
3. Application areas.

Topic

The topic for Futurecasting is the high-level subject or area that the analyst(s) want to investigate. The topic can be broad (e.g., the future of a specific business or market, the

Fig. 2.1 .

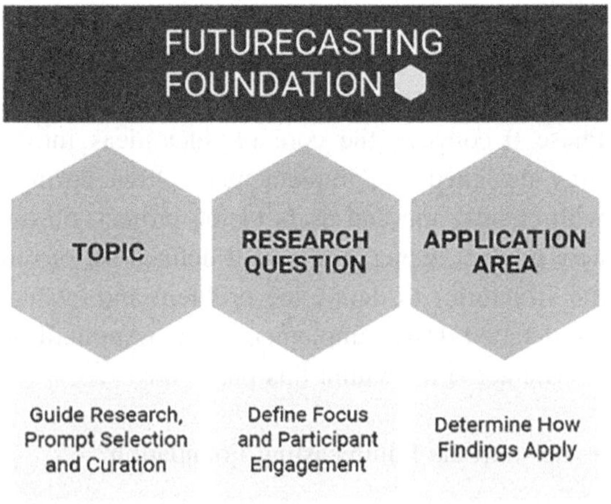

exploration of future products, the effect of climate change on a community, the future of technology and democracy, etc.). It is meant to give the analyst(s) a place to start their broad research and also be able to communicate the nature of the project to others (e.g., additional analyst(s), participant(s), core team, steering committee).

The topic can be determined in multiple ways depending upon the application of the methodology and the desired Application Areas. These Application Areas are the third component of the Futurecasting Foundation; once the analyst(s) has decided upon their topic, and crafted the research question. Applications Areas may need to return to the topic to make sure it sufficiently captures the nature of the project.

At its most simple, the Topic is the area or subject that the analyst(s) want to explore. However, a sponsor or third party who are identified in the Application Areas or who have assigned the topic (e.g., professor, corporate department, government agency) can provide the topic. Wherever the Topic comes from, it is the first building block in the Futurecasting Foundation.

Before moving from a topic to the research question, analyst(s) should gather background research, investigate other explorations of the topic, and explore broader related or ancillary topics. This early research around the topic will help analyst(s) later during curation of the prompts and the participant(s).

Research Question

Once the topic has been selected, the research question aims to narrow the focus. It is important not to try to "boil the ocean." If the research question is too broad, unclear, or not specific enough, the Futurecasting Method will not yield the optimal results. There are a simple set of questions that can help ground the research question (George Mason University Writing Center 2008).

Start by asking explore questions. Ask open-ended "how" and "why" questions about the general topic. Consider the "so what" of the topic. Why does this topic matter? Why should it matter to others? Reflect on the questions that have been considered. Identify one or two questions that are engaging and that could be explored further through research.

Next, determine and evaluate the research question. What aspect of the more general topic.

Do you want the analyst(s) and participant(s) to explore? Is the research question clear? Is the research question focused? Additionally, a research question must be specific enough to be well covered in the resourcing (i.e., time, space) available. Is the research question appropriately complex? Questions shouldn't have a simple yes/no answer and should require research and analysis.

At the Lab we talk about developing a research question that has a constellation around it. Those are elements that you will use to collect your data, so the richer you can start with an exquisite research question, the better your entire project will turn out. How this looks in a question might be something like instead of asking what is the future of climate change? To form a well written constellation around a question you would write it more like "What is the future of climate change as it affects melting polar ice caps and the penguins that live there." Why this is better for the methodology is because it now allows the analyst to look across several sectors: climate, melt rate, polar ice caps, water, penguins.

Sample Research Questions
Appropriately complex:

What are common traits of those suffering from diabetes in America, and how can these commonalities be used to aid the medical community in prevention of the disease?

Asking the Right Question
Ask any scientist or researcher, and they will tell you that finding answers to questions isn't the hard part. If you get the right people together you can generally get the answer to a question. The really difficult thing is to ask the right question. Think of the topic as an aspirational area. The analyst(s) should ask, "What do I want to do? What areas do I want to explore? If I want to see future opportunities, where should I start looking?"

That's the crux of the topic area... it's where the analyst starts. What follows is an investigation on the topic. How can you learn more? What has been said before and who said it? How can your work add to the body of knowledge?

As the analyst(s) researches, they might discover multiple "answers" to the topic. Certainly, they will find multiple perspectives and research.

Next, the analyst should ask, what is my specific question? What is the question I want to ask that has not been asked yet? This will get you to your research question.

The steps for crafting your research question are outlined earlier but a reminder that the analyst needs to be specific: the more specific the question, the more defined the project, the better the outcome. A specific research question will help for all of the following phases of the Method.

When finding the right Question isn't good enough Futurecasting is an applied method. The output is meant to be used by an audience to accomplish a task. This is where simply asking or finding the right question isn't good enough. The analyst has to know who they are asking the question for and what that person or organization will do with the answer.

Application Areas
The Application areas specify how the output from the Futurecasting Method will be used and who will use it. Futurecasting isn't performed in a vacuum. As an applied methodology, the findings outline steps that can be taken by organizations and individuals to take action for their futures.

The Who
The analyst(s) will need to specify who will use the findings.

Example Organizations:

- Academic research;
- Corporations and private industry;
- Government policy makers;
- Military or security professional;
- Trade associations and nonprofits;
- Communities and community organizers.

Each of these organizations will have different requirements and needs to apply the results of the Futurecasting. Understanding who will apply the results will modify how the findings and recommendations are drafted.

The How

The analyst(s) will also need to know how that desired organization is planning on using the results. How it will be applied will modify many of the questions that are asked to generate the raw data and the language used to explore the futures with the participant(s).

Consider the implications and areas the results might be applied.

- What might the results say?
- Why does the argument matter?
- How will the results be put to use?
- How might others challenge the argument?
- What kind of sources will be needed to support the argument?
- What are the budgetary, organizational, stakeholder, and social capital requirements to enable application of the results (think short, medium, and longer term)?

Review and Revise

Based upon the details and specifics of the Application Areas, the analyst(s) may need to modify or revise the topic and/or research question to meet the requirement for impact on a specific group or organization. It is typical to conduct multiple iterations of review on the Futurecasting Foundation to ensure it is solid and will yield the desired output. Recall the goal of the Futurecasting Method is to be applied on hard problems, so continue the revisions until the Foundation is solid. The Who and the How of the application areas can give more detail or specifics to the topic and research question.

This adds a level of complexity to the Futurecasting Foundation that will guide the analyst through the phases but most importantly it will give guidance in the post-analysis phase. Post-analysis can be difficult for the analyst(s) because there are times when they will be "drowning in data."

The Futurecasting Method can generate hundreds of pages of data and in this phase it is up to the analyst(s) to make sense of it. The Futurecasting Foundation will provide clarity and focus for the analyst(s). For this reason, it is important that the analyst(s) continually repeat the process of interrogating the topic, research question, and application areas while developing their Futurecasting Foundation. With a strong foundation, the entire structure of the Futurecasting work will be strong.

2.1.2 Assemble the Core Team, Steering Committee, and Analyst(s)

With the Foundation for the Futurecasting Method in place, the next activity in Phase 0 is to assemble the applicable, broader team. The lead analyst should determine if a core team and a steering committee is needed. Additionally, the lead analyst can determine if other analyst(s) are needed to collaborate through the application of the Futurecasting Method.

Designating a core team, steering committee, and other analyst(s) needed is required regardless of what form of the Futurecasting is being used (e.g., individual, small group, or large group). Generally, the larger the number of participant(s) who will take part in the Futurecasting Workshop, the more necessary a core team, steering committee, or additional analyst(s) become. However, even if an individual Futurecasting Method is being used, it can be helpful to utilize a core team or additional analyst(s) for collaboration, an outside opinion, and/or to validate results.

Core Team

The Core Team for a Futurecasting Workshop is derived from the Applications areas in the Foundation. Understanding who will use the result of the work can give an indication of who could be required to participate in the core team. Generally, the size of the core team is small, from three to five people. Core teams can be made up of people from multiple organizations, domains, and backgrounds. Ultimately, they will be the primary consumers and implementers of the final results.

Activities of the core team include:

- participate in periodic meetings to plan and prepare for the Futurecasting Workshop as well as review the findings;
- review and refine the foundation;
- assist in the identification of prompts and possible communication with SMEs if needed;
- participate in the Futurecasting workshop;
- review and validate the findings;
- socialize and Implement the findings;

Steering Committee

A steering committee is generally used for a large group Futurecasting Workshop. However, they can also be used along with a core team for any form of Futurecasting. As the name suggests, the steering committee's primary purpose is to help steer the effort to success by providing insights. They do not make decisions—that is the core team's responsibility.

The steering committee is larger than the core team and is made up of a wider range of people. The steering committee has fewer activities than the core team but is helpful to broaden the reach and impact of Futurecasting.

The steering committee can include people who will be the main consumers of the Futurecasting findings, but they can also include experts in the Topic or Research Question, secondary or tertiary consumers, as well as a broader set of people to socialize and publicize the results. Because the steering committee should be helpful (and not a distraction), it is important to choose its members carefully.

Activities for the steering committee include:

- review and comment on the foundation;
- recommend prompts;
- make connections to SMEs and participants as needed;
- attend Futurecasting Workshop (optional); and,
- review and comment on findings.

Analyst(s)

The role of analysts in the Futurecasting methodology is the most vital human in the process, guiding it from initial research to actionable outputs through their vision. The Analysts' work ensures the future findings are plausible, and actionable. Here's how analysts contribute at each stage.

Analysts gather diverse research inputs, including social trends, technical data, historical patterns, and expert insights. They select prompts that provoke thinking and align with the goals of the Futurecasting research question and application.

Before or during workshops, analysts facilitate discussions, capture raw data, and translate these insights into synthesized data known as 'prompts' to be used to develop Effect-Based Models (EBM).

After workshops, analysts refine and validate the findings, ensuring they are evidence-based and actionable. They look for cross-cutting themes and confirm the plausibility through additional research or expert consultation.

Analysts write action reports and build presentation decks and sci-fi-prototypes (comic books, graphic novels, animations) to communicate findings effectively. Their work often includes translating complex futures into clear steps for decision-makers to act upon.

Analysts are also often the guides that lead participants through the Futurecasting process, fostering collaboration and innovation.

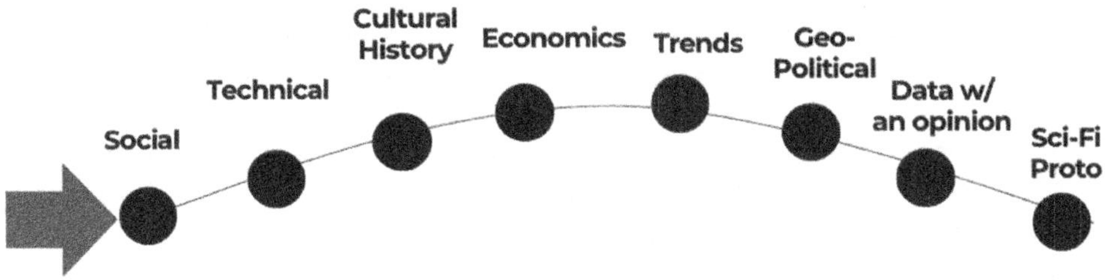

2.1.3 Select and Gather Research Prompts

Prompts are used during the Futurecasting Workshop to inform, inspire, and direct the participant(s). Based upon the Topic and Research question, analyst(s) curate a set of prompts. These prompts can be pulled from a range of interdisciplinary areas outlined in the Futurecasting Framework.

Prompts are represented as the circles on the top arc in Fig. 1.1. These prompts will form the structural foundation of our visions of the future.

Typical prompt categories include:

- social science research;
- technical research;
- cultural history;
- economic/geopolitical projections;
- trends; and,
- data with an opinion.

The prompt categories guide the analyst(s) to gather interdisciplinary perspectives on the topic and research question. Curate prompts from different domains to provide research, perspectives, data, and opinions about the future. This will create the framework for the "art of the possible" on topic areas and give guidance when participant(s) explore the Futurecasting research question.

Social Science Research

Start with a prompt that can help participant(s) understand how society could evolve over the next number of years, that you have chosen for your time horizon, in your topic area. This is because the future, at its heart, is about people. The prompt could capture ideas about social context or how humans will adapt or interact with each other in future. This could also be focused on relationships and connections.

Technical Research

This prompt category can contribute to understanding an interesting technology that is relevant to the Futurecasting topic and research question. Typically, this is a technology that is currently in a Research and Development (R&D) phase within industry or has recently achieved a basic scientific breakthrough. Selecting an early-stage technology allows that technology to be coming into the market and, potentially, enabling a mass effect in ten years—thereby, having a significant influence within the Futurecasting Foundation.

The intersection of these two prompt categories (Social and Technical) will enable an exploration in how this technology will actually be used by future humans. Technology with a long R&D phase rarely is used in the same way as the initial engineers/creators believe it will be used.

Cultural History Research

History is an important prompt for the future. History is the language that people use to talk about and understand the future. Cultural history prompts are used to understand the context for the society or organization that might be affected by the topic or research question.

Additionally, a cultural history prompt can be used to illuminate the impact that the Futurecasting findings might have on the intended application areas. In other words, what is the cultural history of the people and organization that might implement the findings from the Futurecasting? Having an understanding of these populations can be helpful when designating possible actions and outcomes from the post-analysis.

Economic/Geopolitical Research.

The Economic/Geopolitical prompt explores the possible future economic decisions and geopolitical shifts that might affect the people or locations touched by the topic or research question. These could be trends at either the macro- or micro-level depending on the Futurecasting subject. This is important to developing our understanding of the future because economies, geopolitics and their shifting nature will greatly influence how we live and what we can afford. This now builds on our future construct because we are thinking about how individuals are evolving, societies are evolving, technology is evolving, but also how our economy and geopolitical environment is evolving. This then builds into the way of life that people want and the way of life that people can afford. The difference between the two could be an interesting area to explore.

Trends Research

The trend prompt attempts to capture macro or micro trends that could affect the topic or research question. Generally, these trends are regulatory or cultural in nature. For example, governmental policy decisions that affect the topic. Cultural shift or public sentiment can be effective trend prompts. For example, how are the public and the media views of the future of artificial intelligence (AI) and its effect on the future of labor changing?

Data with an Opinion

The Data with an Opinion prompt encapsulates a core idea that drives the Futurecasting Method. People build the future. The future is built by people at organizations. This prompt embraces the biases and opinions of people who are a part of organizations that will have a meaningful effect on the futures that are being modeled.

Prompts

Consider a large auto manufacturing company with a clear stance on the future of cars and mobility—say, they envision a future dominated by electric and autonomous vehicles. This opinion, even if not entirely accurate, becomes a valuable prompt for your Futurecasting because it signals potential actions the company might take, such as investments in EV infrastructure or partnerships with AI firms. These actions could significantly influence the broader landscape of mobility.

Now, imagine juxtaposing this prompt with another one from a competing automaker that instead envisions a future where hydrogen fuel cells and decentralized transportation systems take precedence. The contrast between these two visions creates a dynamic tension in the Futurecasting process, offering a broader spectrum of possible futures.

By including both prompts, we avoid narrowing our focus to a single trajectory and instead embrace the complexity of conflicting viewpoints. These divergent perspectives often highlight gaps, overlaps, or blind spots in the strategies being considered.

Forms and Number of Prompts

Prompts can take many different forms. The form ultimately doesn't matter as long as the prompt serves its purpose. The curation of the prompts across prompt areas when collected together and presented to the participant(s) during the Futurecasting Workshop should give enough data, research, guidance, and inspiration to explore a range of possible and probable futures that address the topic and research question.

Example

For a workshop on urban mobility in 2035, prompts might include:

"By 2030, 60% of urban populations will rely on shared mobility" or "Drones managing transport s of deliveries will likely arrive in major cities by 2035" Data-driven prompts ensure participants explore a range of probable and surprising futures, uncovering both obstacles and opportunities.

Because the prompts are used during the workshop, they will need to be concise so that the participant(s) can understand them quickly and begin to process them. This processing might be simply applying it to the Futurecasting Foundation—exploring the effect the prompt might have on the future. Or the prompts serve as a catalyst for discussion between multiple participants(s). This processing and conversation is captured in the RSW (Research Synthesis Workbook) and used to develop the EBM (Effects based Model.)

The Art of Curating Prompts

There is an art to curating the prompts. It gives the analyst(s) a chance to put their personal stamp on the project. The prompts need to be short but not too simple. They need to be in depth and thought provoking but not so voluminous that they take too much time to comprehend. They should inspire and challenge the participant(s). They should inform participant(s) about emerging technical, social, or trends areas so that the participant(s) are free to explore the futures that answer the research question.

If the prompt, in the collective, informs the participant(s) to answer the research question, then it is acceptable. Likewise, the number of prompts is up to the curation and discretion of the analyst(s). The number should also be dictated by the goal of the prompts. Generally, analyst(s) select three to four prompts to use during the Futurecasting Workshop. This range has been effective because it informs the participant(s) but doesn't overwhelm them with data.

Places to pull your prompts from

- Research papers and articles;
- Business reports and projections;
- Keynote videos and TED talks;
- Interviews;
- Subject matter experts (SME) video;
- Analyst(s) primary research on the topic or research question;
- Podcasts.

Good prompts have enough depth to give the participant(s) something to think about and can be understood in a short period of time. The ideas or research behind these prompts can and often should be complex and subtle but they need to be presented in a way that is functional for the workshop. (Kind of like a short poem!)

Examples of ways to Provide Good Prompts

The following are some examples of ways to provide participants with good prompts.

The analyst(s)' provides a summary:

- Of a 25-page research paper in just a few paragraphs.
- Of a 10-min video about a specific subject in just a paragraph.

- Of a technology demonstration by an expert in just a paragraph.
- Of a long Powerpoint presentation by an expert in just a few sentences.
- Of a Subject Matter Expert Video in just a few sentences.

Examples of ways to Provide Prompts Poorly

The analyst(s)' provides:

- Prompts that are too long in length or too complex for participant(s) to comprehend;
- Prompts can be too vague, short, or unclear;
- A 25-page research paper;
- A Complete book;
- A 45-minute documentary;
- A short bullet pointed list in Powerpoint with vague statements.

Prompts as Categories: "Buckets" or "Containers"

In some instances, the analyst(s) may need more than one piece of research for the prompt to effectively capture the category. An analyst can use a combination of research inputs for a single prompt. Meaning, the analyst(s) might use a video and a research paper to adequately capture or cover the category.

It can be helpful to think of the prompt categories as a "bucket" or "container." The analyst(s) can put one piece of research (e.g., a video) in that container or if needed, they can put two pieces of research in the container (e.g., a video and an article).

This gives the analyst(s) more freedom and creativity to specifically capture the category in the prompt for the participants during the workshop. It is important not to overload the participant(s) with too much information in a single input. They need to be able to comprehend it quickly and begin using it.

Examples

Social Prompt

The analyst wants to explore Gen Z's attitudes toward remote work. They obtain a YouTube short featuring interviews with Gen Z workers discussing their preferences for flexible schedules and work-life balance. This is paired with a report analyzing workplace productivity trends in remote versus hybrid environments to offer a broader understanding of the topic.

Technical Prompt

To investigate the role of blockchain in supply chain security, the analyst obtains a promotional video from a logistics company showcasing their blockchain-enabled tracking system. This is combined with a white paper outlining vulnerabilities and opportunities in current blockchain implementations.

Each of these examples shows how the analyst(s), when needed, can use multiple pieces of research in a single prompt.

Prompts used as Analyst(s) Inputs

Not all of the prompts need to be used during the Futurecasting Workshop. As a part of the research for the Futurecasting Foundation in the specific topic area, these prompts can be used to inform the post-analysis after the workshop. Therefore, these prompts become a part of the raw data that is analyzed.

Example

For a Futurecasting workshop on the future of digital education, the analyst might gather prompts like a news article on AI tutors in classrooms, a podcast discussing teacher shortages, and survey data on student preferences for hybrid learning. While not all prompts are used directly in the workshop, they are retained as raw data to deepen insights during post-workshop analysis.

Subject Matter Experts

Subject Matter Experts (SMEs) interviews are a type of prompt that can be a very effective tool for the analyst(s) when curating a Futurecasting Workshop. These audio or video interviews differ from other prompts in that they are recorded or created by the analyst(s), core team, and/or the steering committee. These interviews are with a current SME on a subject that the analyst(s) would like to use as a prompt.

SME interviews can be effective because they are up to the moment, current research on a specific topic. These interviews can be tailored by the analyst(s) depending on what questions are asked and what topics the analyst(s) requests the SME to cover. However, these videos do take extra effort and time on the part of the analyst(s), core team, and/or steering committee. The following are some practical tips for the analyst(s) when considering a SME for a Futurecasting prompt.

Finding an SME

SMEs are thought leaders who have opinions that the community respects. An effective way to find SMEs is to look at keynote conferences within the domain. There are various websites that highlight the top 10 conferences in a certain domain. Go to those conference websites and take a look at the keynote speakers over the last three to five years. Another idea is to search the TED talk repository for individuals who have spoken on the topic you are looking for. Put in the keep topic as a search on sites like Linked In to identify people who list your topic as their expertise. The key is to avoid initially constraining the search by focusing on people that the analyst(s) or their core team and steering committee can contact. It is important to get thought leaders within the domain as they will be setting the structural foundation of the work and, therefore, will help ensure the work is grounded in science fact.

The Interview

Once the SME has been identified, it is important to understand what the analyst(s) will be asking them in the interview and what you will be asking them to do. The SME will not be a part of the Futurecasting Workshop. This is a very deliberate choice because the SME is a recognized expert on a subject and if they participated then the participant(s) in the workshop might defer to the SME, thus not gathering all the raw data needed for the post-analysis.

The analyst(s) will interview the SME, asking them to record a 5–10-min video clip where they will explore their area of expertise as related to the research question as a prompt directly to the Futurecasting participant(s). The video quality does not need to be high. In fact, it is often optimal if the video doesn't look produced or too flashy. The participant(s) in the workshop will feel they are getting a window into the SME's current research as it relates specifically to the curated area that will be explored.

2.1.4 Select the Participant(s)

> Life's a gloomy puddle, until you start jumping in it.
>
> —Curtis Tyrone Jones

The importance of people remains a cornerstone of the Futurecasting Method.

As important as it is to select a clear topic and research question for the Futurecasting effort, selecting the correct participant(s) for the Workshop is also essential. Participant(s) should be a diverse group of individuals—diversity of thought, experience, background, gender, age and expertise. Part of the group should be thinkers and part of the group doers.

As Futurecasting is an exercise undertaken by practitioners, it is a good idea that they have some domain knowledge. However, participants are also curated who are trained foresight professionals, young participants for fresh and multi-generational perspectives and outliers who come from a variety of backgrounds outside the core domain of the topic. The outliers bring unconventional viewpoints, challenge assumptions, and introduce creative ideas that might not arise from traditional expertise alone. Their unique perspectives can help uncover blind spots, spark innovative solutions, and broaden the range of possible and potential futures explored during the Futurecasting process.

When Futurecasting on corporate topics, participants should represent a diverse mix of roles and perspectives across the organization. This might include individuals from design, engineering, marketing, and operations, as well as leadership from various levels and departments. Additionally, inviting external stakeholders such as key customers, partners, or suppliers can provide valuable insights. However, careful consideration of dynamics is essential. For instance, having a supplier representative and a procurement manager in the same session could foster collaboration (shared goals) or highlight tensions (past negotiations). Thoughtful participant selection seeks tension and ensures a productive and balanced session with actionable outcomes.

Diversity

Diversity of thought is an important part of the Preparation and Curation phase. Bringing in a variety of perspectives helps break through cultural norms that often limit forward-thinking. Whether it's a small family-run business or a global corporation, organizations can fall into patterns of groupthink, making it difficult to envision new possibilities. A mix of contrasting ideas, tensions and differing viewpoints encourages productive debate and pushes participants to explore beyond their usual boundaries, uncovering opportunities and obstacles they might otherwise overlook.

Conflict

A key element of the Futurecasting Method is assembling participants who tackle problems and think in fundamentally different ways, shaped by their diverse and unique experiences. These contrasting perspectives often spark debates or conflicting ideas, which can serve as a catalyst for uncovering innovative and unexpected ideas for future challenges. No single individual or organization holds all the answers, but participants can combine their strengths to navigate uncertainty and discover pathways to opportunities and action.

Bias

Bias is an inevitable factor in the Futurecasting Method, appearing in many forms—cognitive, implicit, unconscious, cultural, and more. Entire fields, from foresight to participatory design, have explored the complexities and impacts of bias. While this textbook can't delve into the full depth of the topic, the Futurecasting Method doesn't avoid bias; it acknowledges and embraces it as a critical aspect of the process.

Analysts should be aware of their own biases and those of the core team, steering committee, participants, and fellow analysts. By confronting and recognizing bias, analysts can work intentionally to mitigate its influence on both interactions and outcomes.

When assembling groups, selecting research, or adding additional analysts, the lead analyst should consider existing biases, understand how they might shape the final results, and identify what perspectives are needed to create more balanced outputs. Bias isn't a flaw to eliminate but a force to manage thoughtfully.

Facilitators

In Futurecasting Workshops, a skilled facilitator is essential. The facilitator guides participants through each step of the process, answering questions, providing direction, and helping to refine or expand their envisioned futures. It's important to note that facilitators are not participants—they have distinct responsibilities and focus on ensuring the process runs smoothly. While the lead analyst often takes on this role, it's not a requirement.

Other key roles support the workshop behind the scenes. Co-facilitators assist during small group sessions, answering questions and keeping teams on track, but they don't lead discussions at the front of the room. Additionally, trained participants—those experienced in the Futurecasting Method—are embedded within small teams. They serve as guides, providing expertise and ensuring the method is applied effectively while encouraging creativity and collaboration.

2.1.5 Draft the Workbooks

Workbooks are tools for capturing the data generated during Futurecasting Workshops, serving as the data repository for post-analysis. They provide participants with a structured space to document their response to prompts, and capture EBMs (Effect-Based Models), the main outputs of the workshop. Two primary types of workbooks are used in the Futurecasting Method: the Research Synthesis Workbook and the Futurecasting Workbook.

These workbooks can range from simple, streamlined data collection tools to more elaborate designs incorporating specific language and visuals. They function as structured databases. Whether in the form of spreadsheets, online forms, or other digital formats, they organize and store the collected information. Analysts rely on these workbooks during the post-analysis process.

Research Synthesis Workbook (RSW)

The RSW is used with the prompts to capture the participant(s) analysis of the curated inputs.

The RSW uses a series of open-ended questions to draw out the perspectives and opinions of the participant(s).

Examples of open-ended Futurecasting questions designed for use in the Research Synthesis Workbook (RSW):

- Understanding Trends:
  What societal, cultural, or technological trends will most influence (your research question/topic area here) in the next 10 years? Why?
- Identifying Opportunities:
  Based on the prompts, what opportunities could arise from this potential future? How might they benefit specific industries or communities?
- Recognizing Obstacles:
  What potential challenges do you see emerging from (your research question/topic area here)? How could they impact key stakeholders?

Futurecasting Workbook (FCW)

The FCW contains a series of questions and tasks that lead the participant(s) through the design and modeling process to develop their EBM. Using the tools of Experience Design, the participant(s) explore a person, in a place, experiencing a problem—all based upon the prompts and the RSW information.

There's a story behind the scenes when you build your EBM workbooks. They can be "forms to fill out" or you, the lead analyst can "plus up" the experience. What's captured in those pages will drive your results, so don't miss the opportunity to inspire the best possible stories!

2.2 Exercise

This exercise will help you develop a research question, collect data, and begin assembling a team for a Futurecasting workshop. This activity helps you practice refining a topic, identifying relevant experts, and designing a diverse group of participants to address your chosen research question.

Part 1: Defining the Research Question

Select a General Topic by starting with a broad area of interest (e.g., future of urban transportation, AI in education, or sustainable energy).

Set Up Your Documentation

Decide how to capture your ideas and organize your work—options include a text document, spreadsheet, or collaboration tools like Google Docs. Think about how to share these materials with collaborators.

Conduct Research

Gather at three or four sources to provide a foundation for your topic. Use academic papers, articles, industry reports, or expert interviews to understand current trends and challenges.

Expand and Narrow Your Topic

Write down related and tangential topics. Explore broadly, then focus on a specific aspect that feels actionable and relevant.

Draft a Research Question

Write a clear, focused, and appropriately complex question. For example, "What are the emerging opportunities for AI in autonomous supply chains by 2035?"

Part 2: Building the Core Team and Steering Committee

Core Team Questions

(a) Who would be helpful to the core team in attending meetings, curating prompts, and socializing findings?
(b) What domains (e.g., academia, government, industry) should they represent?
(c) How would a mix of members make the team stronger?
(d) Are there potential members better suited for the steering committee?

Steering Committee Questions

(a) Do you need a steering committee for connections to experts and participants? Why?
(b) Who should be included, and what domains should they represent?
(c) What resources or networks would they bring to the table?

Part 3: Selecting Analysts

What expertise or perspectives would additional analysts bring?
Are conflicting opinions or different professional backgrounds valuable for the post-analysis?
Identify specific skills or knowledge areas analysts should contribute to the team.

Part 4: Developing Prompt Categories.

Pick four general prompt categories tied to your research question. Examples:

Social trends.
Technological advancements.
Economic/Geopolitical drivers.
Policy changes.

For each category:

(a) Research three thought leaders, noting their names, affiliations, and expertise.
(b) Brainstorm ways to connect with them (e.g., LinkedIn, mutual connections, conference networks).

Part 5: Creating a Diverse Participant List

List individuals working in the field relevant to your research question.

Research their affiliations, projects, co-authors, and areas of expertise.

Use a spreadsheet to track participant details, including:

Domain (e.g., industry, government, academia).
Gender and generation.
Background and expertise.
Links to biographies.

For each participant, find someone with a contrasting perspective to ensure diversity of thought.

Compile a list of individuals with diverse perspectives to participate in the Futurecasting workshop.

Stories from the Lab

A signature style of the Lab includes casual atmospheres, complete with coffee stations, candy at tables, and even a "Flip Flops Encouraged" dress code (because you can't get mad at someone when you can see their toes). We let participants know to check their uniforms, suits and titles at the door. Facilitators have found over and over again that this relaxed vibe helped participants engage fully, encouraging wild ideas and laughter. One participant noted that the comfortable environment made it easier to share unfiltered insights and explore unconventional solutions.

An Interview

Lucy Tournas, J.D. Futurist, Digital Policy Consultant and Strategist, Scholar

I just left a Ph.D. program, and decided to step kind of away from academia after deciding I no longer wanted to be in academia full time. I work for the Center of Industry Self-Regulation (CISR), the research arm of the BBB National Programs, write, occasionally consult, and am a board member of The Institute of Neuroethics (IoNx), a nonprofit think and do tank focused on the development of responsible neuroscience of all. I work on legal, governance, and policy issues.

At the nonprofit, we work with someone who helps us organize public engagement events. And we've been really branching out to kind of teach policy makers, scientists, and developers in the neuro-engineering space how to better think about governance and ethics. For CISR I am helping develop a soft law class, which is something I write about a lot on. I am very interested in thinking about law and governance systems more dynamically, especially in digital technologies. This work helps inform any consulting work I may do with industry, which I hope to build out as there is a lot of opportunity for more internal governance prior to market, strategies that help both industry and society.

One of the projects I wanted to highlight today is this cool neuro-ethics hackathon we developed with the Federation of European Neuroscience Societies (FENS). We did a hackathon with scientists and engineers, but it was quite different than a typical hackathon most scientists and engineers would be used to. But they still were given pieces and asked to build, but instead of a product it was to build a business, thinking about ethics and policy. We blended in elements of futures design and public engagement. We are so lucky to have Darrell Porcello as a lead designer, he is a neuroscientist PhD by education but works is an expert in science education and public engagement. Karen Rommelfanger and Arleen Salles, our Neuroethics were instrumental in coming to him to ask about new ways of teaching ethics. And I am very interested in this engagement as a method of engaging participants with the legal, regulatory, and self-regulatory infrastructure and opportunities they have. I really want them to understand that governance is not something that comes after your design leaves the lab but is built into it. design the governance structure for these things. Often, in the area of neurotech, and especially with the attention Elon Musk has brought it, discussion become very doom or gloom or overhyped. Often discussion get limited to existential treat, omg they're going destroy humanity, or they will save humanity, fix all neuro-diseases, so we can't think about ethics or it will slow it down. I think that is an unproductive wat to think about governance. We need nuance, lots of stakeholders, and methods to think about governing these tools.

For me, one of the reasons why I didn't practice traditional law, is that I really believe these are the interesting questions for our time, governance should be more dynamic. It should be both grounded in reality and able to look forward. For us, this exercise was a power tool in building that bridge between scientists, developers, ethicists, and policymakers, to teach each other how we think. And the participants loved it, instead of a lecture in which they would leave with new information, they left with immediate new skills. Many of the participants are going to be the people who developed the next generation of technology.

We gave them a choice of three product to pitch. One that's pretty close to market, and at this point probably is in market, one that's maybe a year out and one that is more futuristic. And then we had them design their policies, their concerns, their ideas? Once they had time to build, when then added some provocations, such as, what if everyone adopted it, what would a bad actor look like? What would happen if the company lost trust? What would using this product too much look like? There are lots of options, but they highlight a social, economic, ethical, or political change. Many engineers are looking to build a design to fix a problem, and this is a great way to think about externalities.

And in completely good faith. It introduced this idea that ethics and governance is not about limiting you or assuming bad faith development, but that it will leave your lab and it's going to live in the real world with all these other social infrastructures, political infrastructures, legal infrastructures, and what is that going to look like? And knowing that, how does that make you think about technical design or internal policies. It really takes away the tension that can exist between law, ethics and science. And it was amazing to watch the participants think about it in such dynamic way. At the end, they got to pitch their product and policy decisions to a group of expert judges, and the technical and policy solutions they came up with in a few hours were so impressive.

I have spoken and written on the idea of 'Governance by Design; before, which I suggest is the deliberate integration of governance principles, structures, and mechanisms into the development, implementation, and operation of systems, technologies, and organizations. This was watching it in action and in a meaningful way. There is no technology designed without the human and the artifact is in that product. It was just such a joyful way to teach law and ethics. To really do multidisciplinary work, to think proactively in a meaningful way. It was a phenomenal test run for what we could do as a nonprofit, but for me it is also a vision of education go forward, especially in the corporate sector. I want you to think about your technology in future, both all the positive, and to think deeply about economic, political, and social externalities. The point is never to guess them all, or to have a glass ball, but to think more holistically. Industry has a lot of power today, and they need to lead, to learn how to think about governance past financial compliance. Thinking in this manner allows you at least design in a different head space.

I work with all my kids, but my son is currently very into engineering. He learns by building, project-based approaches, so this is layering in ethics and law into those projects. For me it is the only way to teach these concepts. Combining law, technology, business, policy and ethics in a futures approach. These social science concepts are so different than the typical science and engineering curriculum, right? But it could be. And better and more dynamic governance requires it.

Chapter Resources

Vocabulary

Phase 0
The initial "shaping" phase of the Futurecasting Method focused on preparation and curation, setting conditions for success before workshops begin.

Futurecasting Foundation
The structural base of a Futurecasting project, composed of the Topic, Research Question, and Application Areas, which guide all subsequent phases.

Topic
The broad subject or domain chosen for exploration, serving as the starting point for research and communication.

Research Question
A focused, specific, and appropriately complex question derived from the topic that drives the investigation.

Application Areas
The "Who" and "How" of the project—identifying the audience for findings and defining how the results will be applied in decision-making.

Core Team
A small group (typically three to five people) drawn from relevant domains, responsible for planning, refining, and implementing Futurecasting findings.

Steering Committee
A broader advisory group that provides guidance, connections, and feedback without making core decisions.

Analyst(s)
Facilitators and synthesizers of the Futurecasting process who curate prompts, guide sessions, capture data, and transform findings into actionable outputs.

Prompts
Curated research inputs—such as data, trend analysis, expert perspectives, and historical context—that inform and inspire participant exploration during workshops.

Prompt Categories
The six interdisciplinary sources for prompts: Social Science Research, Technical Research, Cultural History Research, Economic/Geopolitical Research, Trends Research, and Data with an Opinion.

SME (Subject Matter Expert)
A recognized authority in a relevant domain who provides targeted, current insight, often through interviews used as prompts.

Research Synthesis Workbook (RSW)
A tool for capturing participant analysis of curated prompts, trends, opportunities, and obstacles.

Futurecasting Workbook (FCW)
A guided workbook used to design and model Effect-Based Models (EBMs) from workshop data.

Effect-Based Model (EBM)
A qualitative representation of a person in a place experiencing a problem, used to explore futures and identify actions.

Bias Management
The process of recognizing and mitigating cognitive, cultural, and organizational biases in participant selection, prompt curation, and analysis.

Governance by Design
The deliberate integration of governance principles, structures, and mechanisms into the development, implementation, and operation of systems and technologies.

Resources

Primary Inputs for Phase 0

- **Social Science Research**—Understanding how societal structures, norms, and human adaptation may evolve.
- **Technical Research**—Emerging or early-stage technologies with potential transformative impacts.
- **Cultural History Research**—Context from past cultural patterns and legacies that shape possible futures.
- **Economic/Geopolitical Research**—Macro- and micro-level forces, policy shifts, and geopolitical changes.
- **Trend Analysis**—Regulatory, cultural, demographic, and market patterns relevant to the topic.
- **Data with an Opinion**—Insightful, opinion-driven sources from influential organizations or thought leaders.
- **Expert Opinions (SMEs)**—Interviews or statements from domain authorities providing specialized, current perspectives.

Recommended Participant Mix

- Outliers with unconventional perspectives.
- Foresight professionals.
- Younger participants for multi-generational perspectives.

- Cross-sector representatives from academia, government, private industry, nonprofits, and community groups.

Useful Links

- Multi-Sided Dice link Dice Shop: https://www.thediceshoponline.com/specialist-dice/multi-sided-dice-sets.html
- Applied Future Lab: https://www.appliedfutureslab.com/
- Threatcasting Community: Threatcasting.ai
- Threatcasting Lab at ASU: https://threatcasting.asu.edu/.

References

AppliedLogix (n.d.) What is a Phase 0 Project for embedded systems development? Retrieved from https://appliedlogix.com/article/what-is-a-phase-0-project

George Mason University Writing Center (2008) How to write a research question. Retrieved from https://writingcenter.gmu.edu/guides/how-to-write-a-research-question

Phase 1 — 3

When you change the way you look at things, the things you look at change.

—Dr. Wayne Dyer

3.1 Research Synthesis

Phase 1 of the Futurecasting Method—Research Synthesis—kicks off with the Futurecasting Workshop, bringing together the planning and preparation from Phase 0. The workshop can vary in size, from a large group to a small team, or even a single individual. It begins with a brief overview of workshop logistics and an introduction to the Futurecasting Method, ensuring participants understand the structure and flow of the session.

Key activities during this phase include:

Presenting the curated prompts to spark ideas and exploration;

- Completing the Research Synthesis Workbook (RSW) to capture thoughts and insights;
- Facilitating discussions to encourage collaboration and diverse perspectives;
- Preparing participants for the transition into Phase 2.

The Research Synthesis phase establishes the foundation for deeper exploration and creative problem-solving, ensuring participants are fully engaged and aligned with the process ahead.

B. D. Johnson and C. Coon, *Futurecasting*, Synthesis Lectures on Threatcasting,
https://doi.org/10.1007/978-3-032-21993-0_3

3.1.1 Presentation of Prompts

Before presenting the prompts, it's important to outline the goals and expected outcomes of both the prompt review and the subsequent Research Synthesis. This core team discussion sets the stage for participants, helping them understand the purpose and direction of their efforts.

The curated prompts serve as a starting point to build the EBM while streamlining the amount of information participants need to process. Throughout the workshop, participants will rely on the prompts provided by the analysts and their own knowledge to envision futures. This approach ensures that the process captures each participant's unique perspectives and opinions, which is critical to the method.

The prompts not only guide the exploration but also introduce participants to new technologies, ideas, and research they might not have previously considered. By balancing structure with open-ended input, the prompts illuminate possibilities while encouraging participants to draw from their experiences and expertise to shape future models.

This phase draws its analytical foundation from the Delphi method, which leverages expert opinion through iterative rounds of questioning and feedback to achieve convergence on key issues (Landeta 2006; Linstone and Turoff 2011).

The Delphi method is a structured approach designed to harness expertise while maintaining anonymity. This avoids distortions often caused by peer pressure in group settings like focus groups. Traditionally, the Delphi process includes three rounds:

- An initial questionnaire asking experts to identify critical issues in a specific knowledge domain (e.g., emerging technologies like metagenomics);
- A second round where experts rate the importance of the identified issues;
- A final round where experts re-evaluate their ratings after reviewing the collective feedback from the group (Birko et al. 2015; Rowe and Wright 1999; Beech 1997; Delbecq et al. 1975).

The Futurecasting Method reimagines the Delphi approach. Instead of seeking consensus, Futurecasting uses curated prompts to provide participants with research, data, and expert opinions on the current state of a domain and its potential evolution over the next decade or other chosen time horizon. The iterative aspect of Delphi is replaced by the unique contributions of participants as they build on prompts with their insights and perspectives.

Rather than narrowing toward a single idea, the goal in Futurecasting is to generate diverse data points that inform the futures framework. These data points are integrated with participant knowledge, fostering a rich synthesis that explores multiple futures instead of converging on one. This method embraces variety and divergence as strengths for robust model development.

3.1.2 Data Capture: Research Synthesis Workbook (RSW)

The primary objective of the Research Synthesis phase is to "tap into the collective expertise in the room." Each participant brings a unique perspective to the topics presented in the prompts, and this phase focuses on capturing those diverse insights.

Participants review the prompts, offering their interpretations, opinions, and perspectives. The aim is to understand the implications of the prompts and identify potential high-level actions that could be taken in response. Documenting this information in the workbook provides valuable insights, including new angles and counterarguments. It helps participants acclimate to the future space they'll engage with more deeply in subsequent phases.

The questions in the Research Synthesis Workbook (RSW) are carefully designed to balance focus and flexibility—guiding participants to explore specific areas while leaving room to capture broader, unexpected insights. This collected data not only informs the next phases of the workshop but also becomes a critical resource for post-analysis and the generation of actionable findings.

Research Synthesis Questions

Please note that these are the general questions typically used upon the participant(s) digesting a prompt. The specific wording is important depending upon the participant(s). The following are the typical RSW questions.

- What was a data point that you found important or interesting?
- What are the implications of the data point on the research questions?
- Are these implications positive, negative, or both?
- What should we do about it?

There are no correct answers to these questions. Each of these questions is purposely designed to be open-ended. "Open-ended questions are used alone or in combination with other interviewing techniques to explore topics in depth, to understand processes, and to identify potential causes of observed correlations. Open-ended questions may produce lists, short answers, or lengthy narratives…" (Weller et al. 2018).

Each question is used to urge the participant(s) to think about the prompt from different perspectives so that these perspectives can be captured in the workbook for later use. As the participants move through the questions, they progressively get more open-ended (Fig. 3.1).

What Data Point Stood Out as Important or Intriguing?

This initial question gives participants an opportunity to jot down their thoughts and impressions, setting the stage for deeper exploration in the following questions. From the perspective of raw data in the post-analysis phase, it helps identify which aspects of

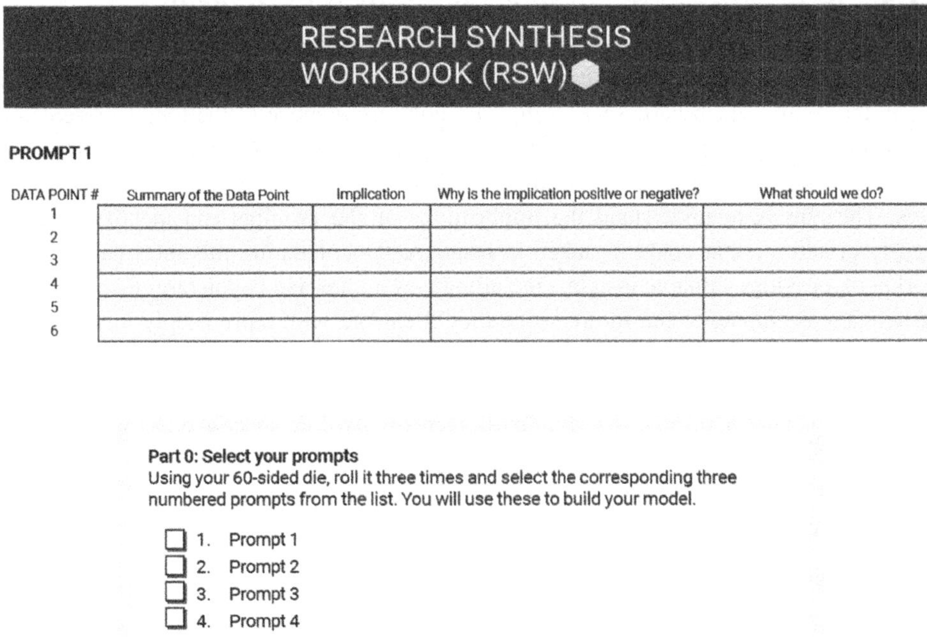

RESEARCH SYNTHESIS WORKBOOK (RSW)

PROMPT 1

DATA POINT #	Summary of the Data Point	Implication	Why is the implication positive or negative?	What should we do?
1				
2				
3				
4				
5				
6				

Part 0: Select your prompts
Using your 60-sided die, roll it three times and select the corresponding three numbered prompts from the list. You will use these to build your model.

- ☐ 1. Prompt 1
- ☐ 2. Prompt 2
- ☐ 3. Prompt 3
- ☐ 4. Prompt 4

Fig. 3.1 Example RSW

the prompts participants found most relevant to the research question. Patterns in their responses can reveal the collective mood or priorities within the group.

How participants describe their selected data point also provides valuable insights. Some may use a broad label, like "AI and Labor," while others might add nuance or commentary, such as "AI-driven labor optimization" or "AI displacing human jobs." These variations can offer clues about each participant's perspective, biases, or priorities, enriching the data set for later phases.

What Are the Implications of This Data Point for the Research Question or Future? This question invites participants to identify what the data point signifies in the context of the research question. It's often one of the most detailed responses, as participants begin connecting the dots between the data and potential futures. This stage encourages participants to explore their biases, opinions, and expertise—core elements of the Futurecasting Method.

For example, a participant might link a data point on autonomous systems with a future where labor markets are reshaped, creating both opportunities for economic growth and threats to worker stability. This step also encourages open expression of contradictions and differing perspectives, recognizing that every participant was selected for their unique contribution to the conversation.

During the post-analysis phase, question two serves as a lens through which the data points are further examined. The responses often expand or refine the original prompts, shedding light on new future implications. These insights don't just add depth to the analysis but also contribute to shaping and informing the futures that will be modeled later in the process.

Are These Implications Positive, Negative, or Both?

Question three encourages participants to consider whether the implications of a data point are positive, negative, or a mix of both. This question also challenges participants to think about for whom the implications apply. By incorporating personal perspectives and considering the experiences of those directly impacted by the potential future, the analysis gains a more human-centered dimension.

Example

A new technological breakthrough, such as an advanced AI system, might be seen as positive for a company that benefits from increased efficiency and profitability. However, it could be negative for workers displaced by automation, highlighting contrasting outcomes depending on the perspective taken.

This question pushes participants to examine the implications from a variety of viewpoints, fostering a richer and more nuanced understanding of the data points. By considering both beneficiaries and those who might be disadvantaged, the analysis captures the complexity of real-world outcomes, providing a balanced foundation for future modeling.

What Should We Do About It?

Question four is deliberately the most open-ended, inviting broad interpretation of the "we." The "we" could refer to the participant(s) themselves, a particular organization, a government, or even humanity as a whole. Participants are encouraged to suggest multiple actions, tailored to different stakeholders or scales.

This question aims to shift the focus from observation to action, prompting participants to think about specific micro-level steps that can address the macro-level challenges posed by the prompts. This type of actionable thinking is critical for later phases of the Futurecasting process, where participants will model potential futures. By thinking proactively here, participants lay the groundwork for practical and strategic insights in the phases to come.

3.1.3 Discussion Aka the "Report Out"

The final activity in Phase 1 focuses on turning the captured data into a collaborative discussion about the prompts. While documenting insights in the Research Synthesis Workbook (RSW) is the primary goal, an equally important objective is sharing and socializing participants' perspectives across the group. This is achieved through a "report out," where participants share their captured data, insights, and opinions with the workshop. This activity concludes the three-step learning and data processing structure of Phase 1.

A tip we like to share at the lab to keep the "Report Outs" short and wrapped up within two minutes is to treat them like a movie trailer. Don't tell everything or it will feel less intriguing and interesting. Kind of like seeing a movie trailer that's too long and you find yourself saying "Now I don't need to even see the movie!" Don't do that. Keep it short and exciting—like a movie trailer!

The "report out" not only deepens individual comprehension but also enriches the collective understanding. Hearing diverse viewpoints helps participants recognize common themes and divergent opinions, sparking further discussion. This dialogue fills the room with valuable data and fosters a shared exploration of the prompts.

Modes of Learning

Phase 1 incorporates these three key learning modes:

- **Passive Learning**: Participants begin by listening to and absorbing the prompts, building foundational knowledge.
- **Active Learning**: Participants then engage in data capture, applying their understanding of the prompts to analyze implications and brainstorm possible actions.
- **Teaching and Collaboration**: The final step involves participants presenting their insights to the group. By teaching others, they solidify their understanding while exposing the group to a variety of perspectives.

3.1.4 Transitioning to Phase 2

The final step of Phase 1 sets the stage for Phase 2, where participants will use the prompts and group-generated insights to create Effect-Based Models (EBMs) for exploring possible and potential futures. Collaborative discussions in this phase ensure that Phase 2 begins with a rich, well-rounded foundation of ideas and perspectives.

At the close of Phase 1, data from each group's Research Synthesis Workbook (RSW) is consolidated into a single master workbook. This combined resource becomes a key input for Phase 2 and serves as a valuable reference for post-analysis.

How Phase 1 Prepared Participants for Futurecasting

- Introduces Key Inputs: Provides participants with relevant research, data, and curated prompts to ground their exploration.
- Focuses Attention: Directs participants toward specific future areas outlined in the research question.
- Builds Connections: Facilitates interaction between participants in the workshop and smaller groups, fostering collaboration and idea-sharing.
- Encourages Exploration: Offers a platform for participants to analyze, discuss, and engage deeply with the prompts.
- Captures Critical Data: Records raw insights and perspectives in the RSW for use in Phase 2 and for post-analysis.
- Shares Perspectives: Familiarizes participants with the viewpoints and interpretations of others, broadening their understanding of the prompts.
- Compiles Inputs: Collects and organizes all raw materials and inputs needed to populate the Futurecasting Workbook.

With these elements in place, Phase 1 ensures that participants are aligned, well-informed, and ready to move forward into the Futurecasting process with confidence and clarity.

3.2 Exercise

Building Collaborative Teams and Effective RSW Design.
Collaboration is critical to maximizing the quality of insights captured in the Research Synthesis Workbook (RSW). Effective teams work together to develop a shared understanding, foster open communication, and balance contributions from all members, ensuring both big talkers and quieter participants contribute meaningfully. This exercise will help you design strategies for building collaborative teams and crafting an effective RSW.

Part 1: Building a Collaborative Team

Task: Construct a checklist for creating a productive and collaborative team. Consider the following questions as you build your checklist:

Team Procedures

What steps or protocols should be in place to ensure the Research Synthesis process progresses smoothly?

Team Agreements

What rules or agreements are essential for maintaining trust, respect, and collaboration within the team? How should these be shared?

Disagreement Management

What strategies or policies should be in place to manage conflicts constructively?

Deliverable: Draft a checklist that includes at least 5 actionable items addressing team structure, communication, and conflict resolution.

Part 2: Identifying Prompts

Task: Create a list of potential sources for prompts in your topic area. Answer the following questions to guide your exploration:

- What are the most urgent or relevant issues within the topic area?
- Who is actively writing, speaking, or researching the topic?
- Who are the thought leaders or experts, and where do they share their insights?
- How might this topic evolve over the next decade?
- Optional: You may use the same topic from Chap. 2 or choose a new one.

Deliverable: Develop a list of at least five sources for prompts, noting why each source is relevant and valuable.

Part 3: Crafting the RSW

Task: Design an RSW that focuses participants on specific future areas while encouraging meaningful discussion. Use the following questions to refine your approach:

- How can the RSW questions focus participants' attention on the key future areas outlined in the research question? How will this influence your phrasing and layout?
- What elements (e.g., structure) could make the RSW more engaging and encourage exploration of the prompts?
- How can facilitators ensure participants are familiar with and consider the perspectives of their peers during the workshop?

Deliverable: Create an outline or sample template for your RSW, ensuring it aligns with your research question and supports collaboration.

By completing this exercise, you'll develop skills in fostering collaboration, sourcing and curating prompts, and designing effective tools for the Research Synthesis process. These are essential elements for maximizing the productivity and quality of your Futurecasting workshop.

Stories from the Lab

During one workshop, the facilitators had carefully prepared the workbooks loaded with a few extra interactive features to test. But as participants started using them, chaos ensued.

One participant exclaimed, "Why does this link ask me only about cat videos?"

It turned out that a rushed late-night edit session left a team member's "place-holder" text. The workbooks became unexpectedly famous for their surprise cat content, and the workshop earned the unofficial tagline: "Thinking about the future? Start with a purr."

The team leaned into the humor, and the accidental distraction actually boosted morale. The takeaway? Always triple-check your work or be ready to embrace the feline future!

An Interview

Dr. Heather Ross Faculty at Arizona State University, Former Special Advisor at City of Phoenix Mayor's Office, Co-Director of SolarSPELL Health, Nurse Practitioner at Honor Health Cardiac Arrhythmia Group, Futurecasting User.

I'm Dr. Heather Ross and I can give you an example of something I'm working on now that has had Futurecasting thoughts woven into it, at least inside my head, the whole time. Years ago when I was embedded in the City of Phoenix in the mayor's office, I came to learn that there was a real problem that our public safety officials were observing: an unfortunately huge and generally growing population of older adults who are living on the street. Our police and firefighters told me more than once that we're seeing all these older folks on the street and we're just sure that they have dementia, but they don't have a formal diagnosis of dementia because they're not well connected to the healthcare system. Police officers and firefighters told me time and again, "These are not people who have strong health and support structures built around them, but we are just sure that they have dementia." But as police officers or as firefighters or crisis responders, our public safety officials were really limited in how to help these older adults with evident cognitive impairment because they don't have the documented diagnosis that would allow them to services that are available for people who are living with dementia.

And so, with that nugget in mind, I went about imagining how can we address this problem and what might a future look like where older adults who are living on the street might not even *be* on the street if it was dementia that got them there. Or they might be able to get the services that they need to navigate their way out of homelessness, that right now they're not able to access because they're trying to get through their day with dementia or other cognitive impairment. Today, we just don't have systems set up around that.

And so, in trying to imagine this future where older adults with dementia or other cognitive impairment don't fall through those cracks, we went about thinking through the processes of how to change that system— the whole Futurecasting process, where you imagine what this desired future looks like, and then you work backward, through what are the steps, what are the gates that are going to get us there? Where are the threats that are going to come up and poke a hole in the bubble of your plan? Because I came to

Futurecasting through Threatcasting for many years, that approach was ingrained as part of the way that I've been thinking about how to get at this problem. At this moment I still do not have the luxury of a whole bunch of directed funding to tackle this problem, and so I have had to be quite entrepreneurial with little small pots of opportunity that present themselves and a couple of small grants that have been enabling. But to date, we have been able to establish dementia screening in homeless shelter settings, and we've been able to establish a pipeline of referral and initial diagnosis for people experiencing homelessness who have a positive screen for dementia in healthcare for the homeless clinic setting.

I knew that we needed all of these systems in place, and I know that ultimately, we need even more upstream screening systems in place before people experience homelessness. But it was thinking through what those opportunities are, what those threats are, and keeping that desired future state in mind that has absolutely enabled this work. While we remain in the very early phases of operationalizing our plans, we continue to modify as life throws us really interesting side paths, curve balls and sometimes tomatoes getting thrown at us, which I choose to make like tomato sauce out of! But I can tell you that today, there are senior adults who find themselves in a homeless shelter and within one week are found to have some cognitive impairment and are seeing a specialized healthcare provider who is able to identify that they have dementia and get them hooked up with the medications and longer-term services that they need to help slow down that progress of dementia.

The Threatcasting approach, and turning it on its side to the Futurecasting approach, was so ingrained for me that I can't say that I was conscious that that's how I was thinking about it until I was doing it and realized, "Oh, I have this desired future state in mind. I've thought backwards through what are the steps that we need to take, and I've of anticipated what some of those threats are gonna be, and already thought how to side run around them."

Chapter Resources

Vocabulary

Phase 1—Research Synthesis
The first active workshop phase of the Futurecasting Method, where curated prompts are presented, participants capture insights in the Research Synthesis Workbook (RSW), and collaborative discussions prepare for modeling in Phase 2.

Prompt Presentation
The structured introduction of curated prompts that provide participants with data, trends, and expert perspectives to inspire exploration and scenario building.

Research Synthesis Workbook (RSW)
A structured tool for recording participant interpretations, implications, and potential actions related to each prompt.

Delphi Method
A structured expert-elicitation process using iterative rounds to converge on key issues; adapted in Futurecasting to collect diverse perspectives rather than consensus.

Implication Analysis
The process of determining how a given data point might affect the research question or future scenarios, including positive, negative, and mixed outcomes.

Action Recommendations
Open-ended participant suggestions for what specific stakeholders ("we") should do in response to the prompt's implications.

Report Out
The group-sharing stage in which participants present key insights from their RSW entries to the larger workshop, fostering collective understanding.

Modes of Learning in Phase 1

- **Passive Learning**—Listening and absorbing information from prompts.
- **Active Learning**—Capturing and analyzing implications in the RSW.
- **Teaching and Collaboration**—Sharing insights with the group to broaden perspectives.

Master Workbook
A consolidated version of all participant RSW entries, serving as a critical input for Phase 2 and post-analysis.

Primary Inputs for Phase 1

- **Curated Prompts**—Drawn from interdisciplinary sources such as social science research, technical research, cultural history, economic/geopolitical trends, and SME interviews.
- **Participant Expertise**—Knowledge, perspectives, and lived experience from a diverse participant mix.
- **Facilitation Frameworks**—Structured guidance for presenting prompts, managing time, and encouraging balanced participation.

Prompt Sources

- Academic research papers and reports.
- Industry forecasts and white papers.
- Policy briefs and regulatory analyses.
- Cultural studies and historical accounts.
- SME interviews and keynote presentations.
- Trend and market analysis reports.

Recommendations

1. **Set Clear Expectations Early**—Outline Phase 1 goals before presenting prompts to focus participant attention.
2. **Balance Structure with Freedom**—Use open-ended RSW questions that guide thinking without restricting creativity.
3. **Embrace Divergence**—Value differing interpretations and contradictions as inputs for richer future models.
4. **Encourage Cross-Pollination**—Use the "report out" to expose participants to others' perspectives.
5. **Capture Raw Data Thoughtfully**—Ensure all RSW entries are legible, complete, and consolidated for Phase 2.
6. **Manage Time Rigorously**—Use "movie trailer" pacing for report outs to maintain energy and engagement.

Useful Links

- Multi-Sided Dice link Dice Shop: https://www.thediceshoponline.com/specialist-dice/multi-sided-dice-sets.html
- Applied Future Lab: https://www.appliedfutureslab.com/
- Threatcasting Community: Threatcasting.ai
- Threatcasting Lab at ASU: https://threatcasting.asu.edu/.

References

Beech B (1997) Studying the future: a Delphi survey of how multi-disciplinary clinical staff view the likely development of two community mental health centres over the course of the next two years. J Adv Nurs 25(2):331–338. https://doi.org/10.1046/j.1365-2648.1997.1997025331.x

Birko S, Dove ES, Ozdemir V (2015) A delphi technology foresight study: mapping social construction of scientific evidence on metagenomics tests for water safety. PloS One 10(6):e0129706. https://doi.org/10.1371/journal.pone.0129706

Delbecq AL, Van de Ven AH, Gustafson DH (1975) Group techniques for program planning: a guide to nominal group and delphi processes. Scott, Foresman

Landeta J (2006) Current validity of the Delphi method in social sciences. Technol Forecast Soc Change 73(5):467–482. https://doi.org/10.1016/j.techfore.2005.09.002

Linstone HA, Turoff M (2011) Delphi: a brief look backward and forward. Technol Forecast Soc Change 78(9):1712–1719. https://doi.org/10.1016/j.techfore.2010.09.011

Rowe G, Wright G (1999) The Delphi technique as a forecasting tool: issues and analysis. Int J Forecast 15(4):353–375. https://doi.org/10.1016/S0169-2070(99)00018-7

Weller SC, Vickers B, Bernard HR, Blackburn AM, Borgatti S, Gravlee CC, Johnson JC (2018) Open-ended interview questions and saturation. PloS One 13(6):e0198606. https://doi.org/10.1371/journal.pone.0198606

4 Phase 2

Imagination is more important than knowledge.

—Albert Einstein

4.1 Futurecasting

Transitioning to the Futurecasting Phase

In the next phase of the Futurecasting Method, participants use the insights and data captured in the RSW to imagine possible and potential futures. This phase marks a shift from the broad, high-level perspectives provided by the curated prompts to the creation of a focused and detailed Effect-Based Model (EBM).

Participants construct their EBM by combining elements of participatory design, experience design, and the science fiction prototyping process. The phase concludes with the development of a detailed future.

Key Activities in This Phase:

- Develop EBMs: Participants create detailed models of futures, incorporating data, insights, and creative thinking.
- Capture Raw Data: All supporting data, perspectives, and ideas generated during this phase are recorded for post-analysis and future use.

B. D. Johnson and C. Coon, *Futurecasting*, Synthesis Lectures on Threatcasting,
https://doi.org/10.1007/978-3-032-21993-0_4

This phase builds a bridge between conceptual exploration and actionable foresight, setting the stage for practical strategies to address the envisioned futures.

4.1.1 Effect-Based Model (EBM)

The EBM is the central tool in the Futurecasting Method. It is a qualitative model that enables participants to use insights from the prompts and the RSW to envision possible and potential futures.

Typically, a 10-year time horizon is employed for Futurecasting, striking a balance between imaginative exploration and grounded foresight. A decade is far enough ahead to extend beyond typical product cycles, political terms, or career spans, allowing participants to consider a broader range of future possibilities. At the same time, it remains close enough for realism: technologies likely to influence the future already exist, are being tested in labs, or are in the early stages of development. This timeline allows participants to think creatively while staying anchored to plausible developments.

At its simplest, the EBM is a structured framework—often a database or spreadsheet—that poses targeted questions and guides participants through specific tasks to construct a qualitative model. The process transitions from macro-level exploration of broad research and perspectives (curated prompts and the RSW) to a micro-level focus on a particular future. Participants explore how an individual in a specific context might experience the identified future.

This person-centered, detailed approach differentiates the EBM from other foresight tools like scenario planning (Van der Heijden 2011) or the futures wheel (Glenn 1972). By narrowing down to actionable, human-centric insights, the EBM provides a unique and practical way to model futures with depth and precision. While many foresight tools focus on identifying and describing future scenarios, Futurecasting places equal emphasis on actionable outcomes. The process doesn't end with envisioning a potential future—it continues with backcasting. Example: Unlike traditional scenario planning, which might conclude with a descriptive narrative of a dystopian or utopian future, Futurecasting goes further by mapping concrete steps organizations can take to avoid or intentionally enable that future.

Other Elements Unique to Futurecasting

Participatory Design

Using Participatory Design to Build the EBM the Futurecasting Method leverages participatory design to capture rich, detailed data in the Effect-Based Model (EBM). Participatory design centers on the direct involvement of people in co-creating tools, environments, systems, and solutions. According to Robertson and Simonsen (2012):

> At the core of Participatory Design is the direct involvement of people in the co-design of tools, products, environments, businesses, and social institutions. Participatory Design has developed a diverse collection of principles and practices to encourage and support this direct involvement. Many of the design tools and techniques generated to further this process have become standard practice for the design and development of information and communications technologies and increasingly other kinds of products and services. These tools include workshops where participants collaboratively envision future practices and products, as well as scenarios and personas that empower people to represent their own activities and ideas.

This collaborative approach is a cornerstone of Futurecasting. It encourages discussion, debate, and the sharing of diverse perspectives, resulting in more detailed and nuanced data for the EBM. Unlike methods that rely heavily on predefined inputs, such as the futures wheel (Glenn 1972), Futurecasting uses curated prompts to spark exploration. Participants actively shape future models based on their unique experiences and viewpoints, making the process dynamic and inclusive.

Experience Design

The questions and tasks within the Effect-Based Model (EBM) draw from a range of tools, theories, and practices, with a strong influence from experience design. In Futurecasting, participants are guided through tasks and questions designed to explore how future opportunities might impact a specific person in a specific location. This experience-based approach shifts the mindset of participants, encouraging them to think deeply about potential futures and their effects on human lives.

Experience Design emerged in the early twenty-first century, originally rooted in the high-tech and information technology (IT) industries. It began with human–computer interaction (HCI), focusing on how users engaged with computers and how systems responded to those interactions (Karray et al. 2008). This evolved into User Experience (UX), which refined the measurable and desired outcomes of product interactions. Experience Design expanded these principles, applying them broadly across products, services, and systems.

As technology became more integrated into daily life, Experience Design sought to document, explain, and design interactions beyond the software or device itself. It encompassed the entire user journey, including setup processes, advertising, marketing, and ongoing services.

Shedroff (2001), in Experience Design, emphasized that the most important aspect of any design is how it is understood in the mind of the audience. He introduced the idea of using cognitive models and metaphors to help people recognize and understand concepts in new ways. "New cognitive models can often revolutionize an audience's understanding of data, information, or an experience by helping them see things they previously couldn't," Shedroff wrote. He later expanded this idea, saying, "Every sketch, model, and prototype is an elaborate fiction on the road to becoming something real" (MacWorld Stage 2011).

Futurecasting adopts these principles to shape the EBM process. Participants are not tasked with defining a single opportunity or predicting a specific outcome. Instead, they are guided through a series of questions designed to capture the effects of future opportunities on individuals, communities, or systems. This approach enables participants to consider complex interactions and second- and third-order effects, uncovering a richer, more detailed vision of the future.

By using an experience-based framework, Futurecasting ensures that the EBM captures the human dimension of potential futures, making the insights more tangible, relatable, and actionable.

Science Fiction Prototyping (SFP)

Using Science Fiction Prototyping (SFP) to Envision Opportunities

The Futurecasting Method leverages the Science Fiction Prototyping (SFP) process to expand participants' ability to model and imagine future opportunities. Through a series of targeted questions, participants draw on insights from the prompts and the RSW to populate the EBM with rich, detailed data about possible futures (Fig. 4.1).

Science Fiction Prototyping integrates creative disciplines—such as storytelling, comics, film, and art installations—into the process of innovation. It enables participants to invent and explore potential futures using trends and insights from research and Futurecasting as a foundation.

SFP offers a unique approach to examining the human, societal, and systemic impacts of potential opportunities. These impacts can include ethical, cultural, legal, or business considerations. The methodology is rooted in a straightforward five-step process:

Five-Step SFP Process

1. **Build the World**

- Define a specific setting.
- Who is the main character?
- Where do they live?
- When does this take place?
- Choose a timeframe just beyond what feels plausible.

2. **Insert a Technology Inflection Point**

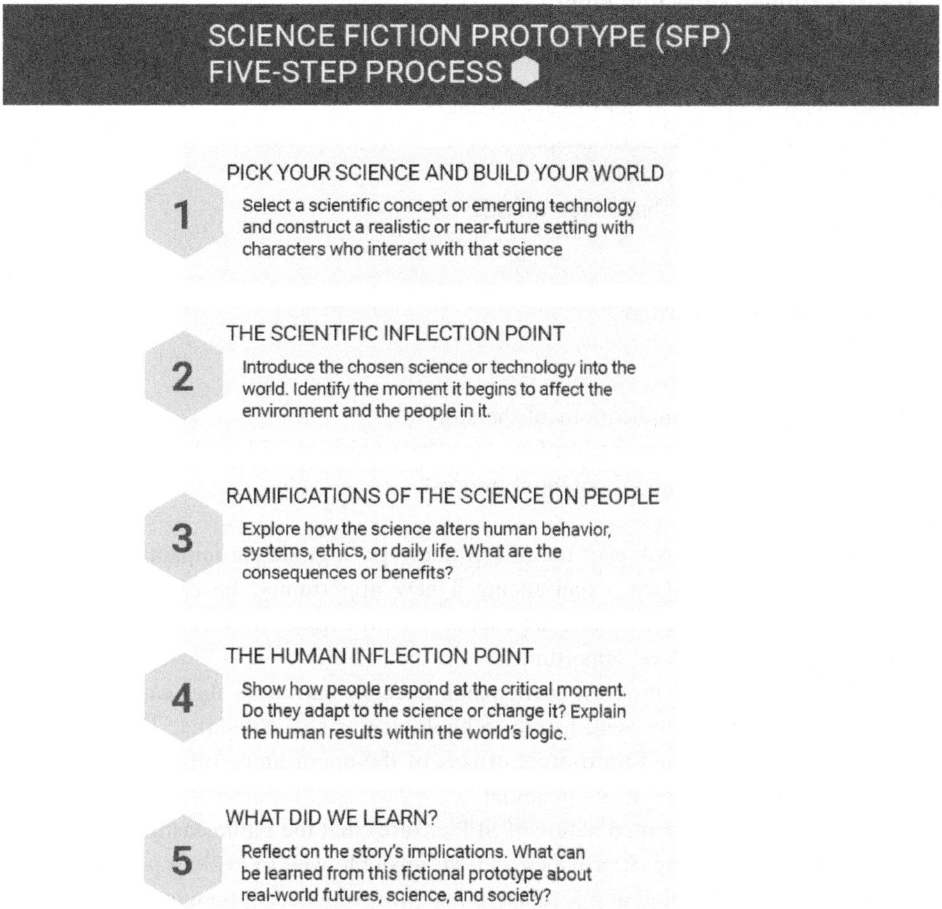

Fig. 4.1 Science Fiction Prototype (SFP) five-step process (Johnson 2011)

- Introduce a technology or innovation that excites or intrigues you.
- What change does it bring to this world?

3. **Explore the Ramifications**

- Examine the consequences of this technological change.
- How does it affect the characters, their environment, and their daily lives?

4. **Insert a Human Inflection Point**

- Explore how people respond to these changes.
- How does the main character adapt?
- What decisions do they make?
- How do those choices shape their world?

5. **Reveal What We Learned**

- Reflect on the larger implications of the story.
- What insights can be drawn from how the opportunity played out?
- What does it teach us about potential futures?

The Futurecasting Workbook uses the SFP framework to guide participants through envisioning a person, in a place, experiencing a new opportunity. Instead of focusing on challenges or obstacles, participants are encouraged to explore how individuals or groups adapt to and benefit from the opportunity.

By following the SFP process, participants develop narratives that capture complex, human-centered interactions with future technologies or societal shifts. These narratives help to uncover second- and third-order effects of the opportunity, offering a deeper and more holistic understanding of its potential.

The creative and experiential nature of SFP ensures that the Futurecasting Method goes beyond abstract forecasting. It provides a vivid, relatable way to explore possible futures, fostering innovative thinking and actionable insights grounded in human experience.

Effects-Based Operations

After opportunities are described in the workbook, the questions and tasks shift to explore what factors might bring about or enable those opportunities. This process draws on the military concept of Effects-Based Operations (EBO), adapted here to a future-focused, opportunity-driven context.

EBO is defined as "a process for obtaining a desired strategic outcome or 'effect' through the synergistic, multiplicative, and cumulative application of the full range of military and nonmilitary capabilities at the tactical, operational, and strategic levels" (Rickerman 2003). Instead of focusing on adversaries, the concept is reimagined in Futurecasting to consider how desired outcomes might emerge in a complex, interconnected future.

In foresight, this aligns with causal layered analysis (CLA), which examines futures through multiple layers of understanding to uncover deeper drivers and possibilities. CLA moves beyond surface-level trends or immediate consequences, exploring the systems, worldviews, and myths that underpin them. By addressing these deeper layers, practitioners can identify not only the potential impacts of an opportunity but also the shifts needed to enable or amplify it (Inayatullah 1998).

Exploring Opportunities from Multiple Angles

Participants are guided to think about opportunities through a 360° view, considering how social, technological, cultural, economic, and environmental factors interact to enable those opportunities. This whole-of-society approach expands the scope of what can be modeled, encouraging participants to imagine a broader range of possible futures while generating richer, more comprehensive data.

By examining opportunities in this way, Futurecasting doesn't just describe possible outcomes but actively maps how they might come into being. This creates a more dynamic and actionable framework, allowing participants to think critically about how to shape or respond to emerging possibilities.

4.1.2 Capture the Data

This section outlines each step in the Futurecasting phase, directly tied to the tasks and questions in the Futurecasting Workbooks. These tasks form the core of the Effect-Based Model (EBM) that participants will construct. Sample workbook visuals are included to help clarify these concepts (Fig. 4.2).

Selecting Foundational Data Points

The prompts and their corresponding insights from the Research Synthesis Workbook (RSW) serve as the raw materials for creating plausible, opportunity-focused futures rooted in current research and science.

The first section of the Futurecasting Workbook requires participants to identify foundational data points curated from Phase 1 prompts. Participants review the RSW commentary and select one key data point from each prompt they were presented. These selections will inform the rest of their Futurecasting exercise.

Quick Selection for Efficient Progress

To maximize time spent on building the workbook, participants are encouraged to select their data points quickly, without overanalyzing. Facilitators can employ techniques like rolling dice, drawing cards, or using random number generators to introduce an element of randomness, which can lead to surprising combinations. Contradictory prompts, for

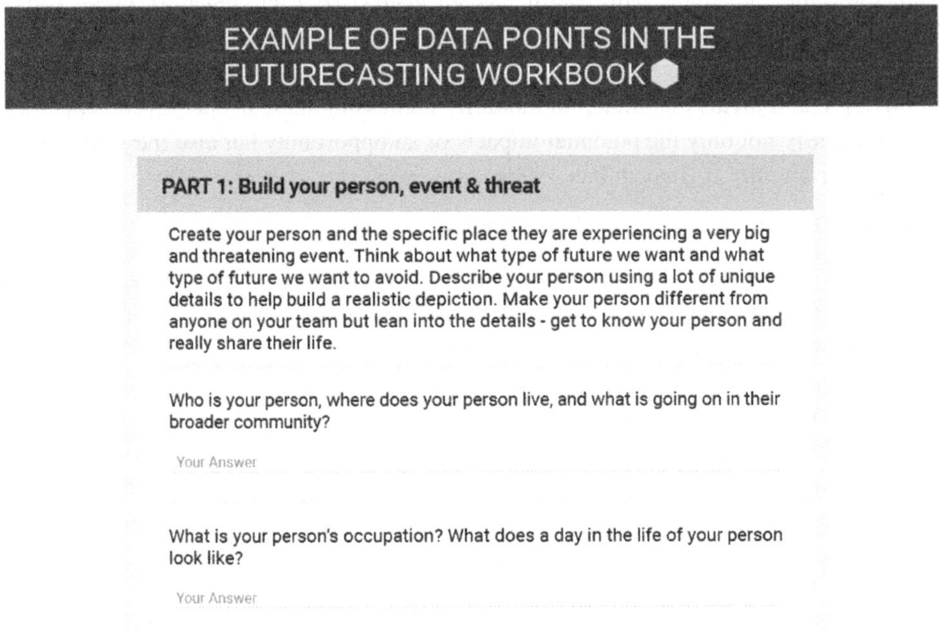

EXAMPLE OF DATA POINTS IN THE FUTURECASTING WORKBOOK

PART 1: Build your person, event & threat

Create your person and the specific place they are experiencing a very big and threatening event. Think about what type of future we want and what type of future we want to avoid. Describe your person using a lot of unique details to help build a realistic depiction. Make your person different from anyone on your team but lean into the details - get to know your person and really share their life.

Who is your person, where does your person live, and what is going on in their broader community?

Your Answer

What is your person's occupation? What does a day in the life of your person look like?

Your Answer

Fig. 4.2 .

example, may push participants to reconcile differences creatively, resulting in richer and more dynamic futures.

Building the EBM: A Person in a Place, Experiencing a Problem That Is Solved

After grounding their exploration in the selected data points, participants begin to imagine a specific person living in the envisioned future. This human-centered approach adds depth and tangibility to the EBM by focusing on individual experiences rather than abstract systems.

Humanizing the Future

Participants are tasked with building a character in their workbooks. Participants are encouraged that they are writing a movie script and they should give as many details as possible to build their character. They are asked to consider:

- Who is this person? Define their identity, background, and role in the community.
- Where do they live? Visualize their environment—whether physical or digital.
- What is their community? Every person is connected to a broader network, whether family, colleagues, or an online group. This community provides support and shapes the person's experience.

By narrowing the focus to one individual, participants can avoid feeling overwhelmed by the complexity of imagining large-scale changes. Instead, they create a relatable narrative that highlights the human dimension of potential futures.

Details Make the Future Real

Depending on the audience and research goals, participants can flesh out intricate details about the individual's daily life, environment, aspirations, and challenges. For example:

- What are their hopes and dreams in future?
- How do they interact with the opportunities presented by the selected data points?
- How does their environment shape their ability to adapt or thrive?

Describing the Event
The next step asks participants to describe the Problem that is Solved or the Opportunity—the moment or process through which the opportunity takes form. This could be a physical development, a digital innovation, or a societal shift that profoundly impacts the individual's life.

To ensure a rich exploration, participants answer a series of questions designed to push them toward detailed, imaginative, and plausible scenarios. The goal is to capture not only the immediate effects of the opportunity but also its broader implications on the person, their community, and their environment. Examples of questions could include:

- What immediate changes does the solution or opportunity bring to the individual's daily life, and how do they adapt to or embrace these changes?
- How does the solution or opportunity influence the broader community, both positively and negatively, and what new interactions or dynamics emerge as a result?
- What are the longer-term impacts of this solution or opportunity on the person's environment, such as economic, social, or technological shifts, and how do these changes shape their future?

4.1.3 Taking the Data Deeper

Futurecasting moves beyond the basic "5Ws" (who, what, when, where, why) of traditional information gathering to craft deeper, more nuanced narratives. The questions are designed to explore the complexity and uncertainty of future opportunities and the environments in which they may emerge. These questions encourage participants to imagine detailed, human-centered scenarios that illustrate how opportunities unfold and affect individuals and communities.

Example Questions

- Who is your person, and what is their broader community?
- What is their occupation? What does a typical day in their life look like in this future?
- Where do they live? What does their environment (physical or digital) look like?
- What is the problem to be solved that introduces the opportunity into their life?
- Imagine your person engaging with the moment the problem is solved or the opportunity is first experienced:
  - What is happening?
  - Who else in their life is involved?
  - What positive outcomes might the opportunity create?
  - How does the opportunity address or amplify their hopes, aspirations, or challenges?

By answering these types of questions, participants build a vivid and relatable picture of the future, highlighting how the solution to the problem or the opportunities interact with people, communities, and environments in dynamic and meaningful ways.

Experience Questions

The next section of the Futurecasting Workbook draws heavily from the principles of experience design, guiding participants to craft a deeper, richer understanding of how opportunities might unfold for individuals and communities. Participants are asked a series of curated questions to expand their descriptions of the person's experience with the solution or opportunity and the event that introduces it into their lives (Fig. 4.3).

These questions are designed not only to gather the data necessary to address the research question but also to challenge participants to expand their thinking and imagine the future from multiple perspectives. Unlike earlier questions that focused on open-ended exploration of the person and their environment, these experience questions probe participants to envision how the opportunity would feel, look, and play out in real time from the perspective of the person they've imagined.

Why Experience Questions Matter

Experience questions help participants dive into the lived experience of an individual engaging with the opportunity, fostering a more detailed and empathetic narrative. By focusing on the experience itself, participants uncover nuances and second- or third-order effects that might not emerge in a more abstract exploration. These details feed directly into subsequent stages of the Futurecasting process, such as exploring enabling factors and future implications.

If time or group size is a constraint, facilitators can prioritize a subset of experience questions to ensure the process remains efficient while still generating meaningful insights.

Example Experience Questions

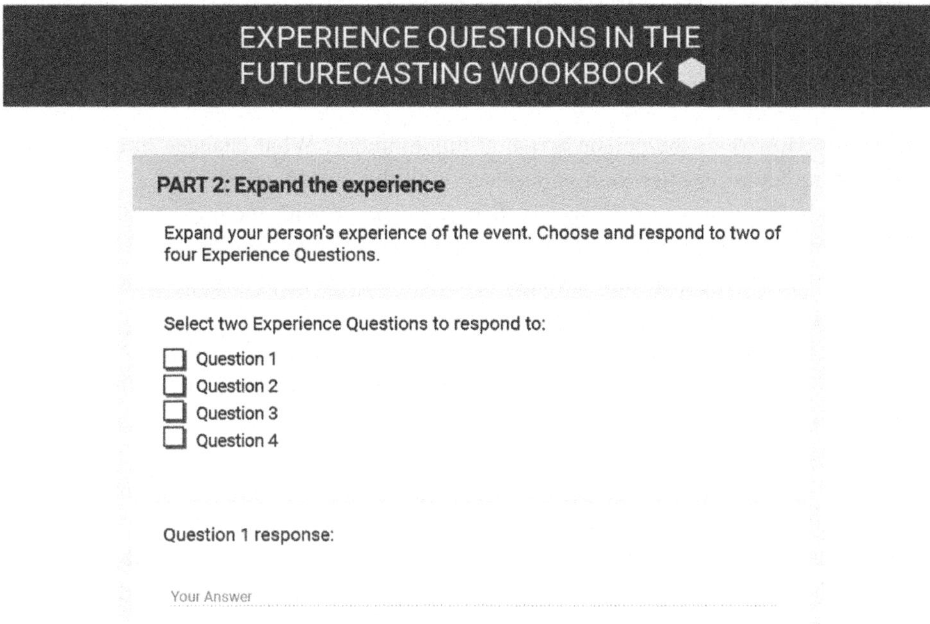
EXPERIENCE QUESTIONS IN THE FUTURECASTING WOOKBOOK

PART 2: Expand the experience

Expand your person's experience of the event. Choose and respond to two of four Experience Questions.

Select two Experience Questions to respond to:

- [ ] Question 1
- [ ] Question 2
- [ ] Question 3
- [ ] Question 4

Question 1 response:

Your Answer

Fig. 4.3 Example of experience questions in the futurecasting workbook

- How does the solution or opportunity initially present itself to the person?
- What do they see, hear, or feel when they first encounter it?
- What do they misunderstand or fail to realize until later?
- How does the person engage with the solution or opportunity in their daily life?
- What new actions, routines, or interactions are introduced?
- How does it enhance or challenge their usual activities?
- How does the person connect with their community through the solution or opportunity?
- Where and how do they communicate with others about it?
- What new connections or shared experiences arise as a result?
- What new capabilities does the solution or opportunity provide to the person and their community?
- How do these capabilities improve their lives or create new possibilities?
- What ripple effects does the solution or opportunity create beyond the person's immediate experience?
- How does it influence their community or broader environment over time?
- What would an ideal outcome look like for the person and their community?
- How does the solution or opportunity help them achieve their goals or dreams?

Designing the Experience—Putting It into Practice

Experience design asks participants to explore the journey their imagined person takes before, during, and after engaging with the opportunity. This includes:

- Duration: How does the person arrive at this moment? What changes in their state of mind from before to after their experience with the opportunity?
- Interaction: How active or passive is their engagement with the opportunity? Are they fully immersed, or does it become a seamless part of their routine?
- Intensity: Does the experience demand conscious engagement (e.g., learning something new), or does it evolve into a habit over time?
- Design Elements: What does the experience look, sound, and feel like? How do design choices (e.g., colors, sounds, interfaces) enhance the person's interaction with the opportunity?
- Value Creation: The ultimate purpose of the opportunity is to provide value to the person and their community. Value can be explored across five dimensions:
  - Functional Value: Does it solve a problem or make their life easier?
  - Financial Value: Does it create economic opportunities for them or others?
  - Emotional Value: How does it make them feel (e.g., inspired, connected, nostalgic)?
  - Identity Value: Does it foster a sense of belonging to a larger group or community?
  - Meaningful Value: Does it represent something greater than itself, contributing to a sense of purpose?

Using Experience Design to Inform the Future

By focusing on these six dimensions—duration, interaction, intensity, design, value, and community—participants create a vivid and holistic picture of how opportunities might be experienced in future. This process enriches the Futurecasting Model with actionable insights, ensuring that the envisioned opportunities are grounded in human-centric narratives and meaningful impacts.

Experience design bridges imagination and reality, helping practitioners shape futures that are not only plausible but also relatable and inspiring.

Enabling Questions

As with all great speculative narratives, Futurecasting encourages a shift in perspective to fully understand the broader context of the "event" and the forces that enable it. Instead of focusing solely on the individual experiencing the opportunity, participants step back to explore the systems, actors, and conditions that made the opportunity possible (Fig. 4.4).

Participants are prompted to think about the motivations, decisions, and circumstances that allowed the solution or opportunity to emerge. These enabling questions are designed to challenge assumptions and push participants to consider perspectives they might not have initially explored. For example, what actors or systems drove the creation of the

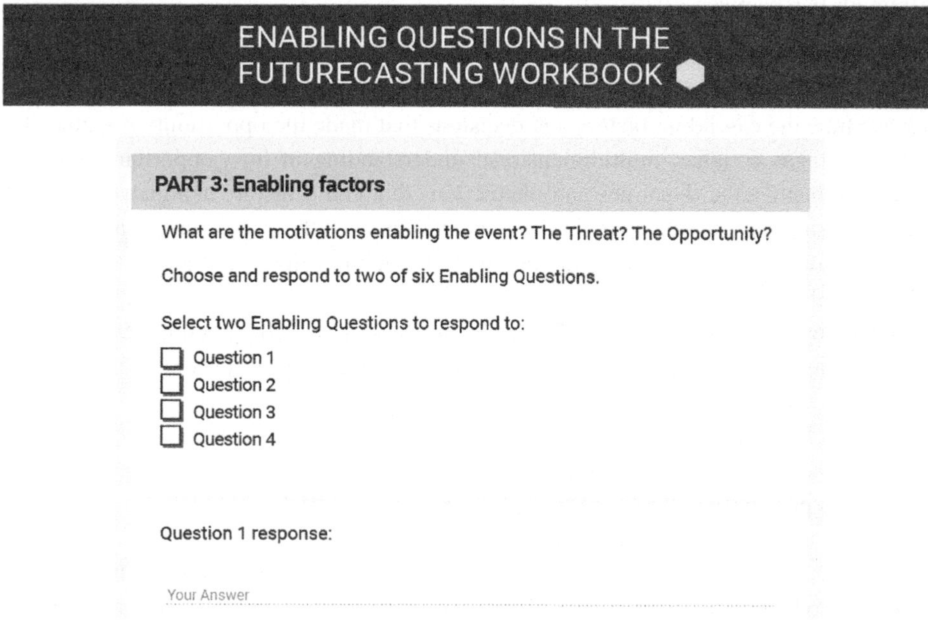

Fig. 4.4 Example of enabling questions in the futurecasting workbook

solution or opportunity? How might their intentions have evolved over time, and what broader dynamics supported their actions?

This phase encourages participants to move beyond a single perspective and consider the solution or opportunity at a macro level, leveraging their own expertise, experience, and even biases to imagine enabling factors. While some of this strategic-level information may have been touched on during the Research Synthesis phase, this stage focuses on tactical and practical elements that bring the opportunity to life.

Example Enabling Questions

- Who or what enabled this solution or opportunity to come into being?
- What systems, innovations, or societal shifts played a role?
- Were there key individuals, organizations, or movements that contributed?
- What motivations drove these actors or systems to act?
- Were they pursuing specific goals or responding to external pressures?
- How might their initial intentions have shifted over time?
- What conditions made this solution or opportunity possible?
- What cultural, economic, or technological trends are aligned to support it?
- Were there unintended consequences or unforeseen collaborations that amplified its impact?

Reframing Perspectives

This phase of Futurecasting requires participants to set aside the individual perspective they developed earlier and adopt a broader, more objective view. Thinking systemically, they examine the cascading factors and decisions that made the opportunity possible. This shift encourages a richer, multidimensional understanding of how opportunities unfold, helping to identify the dynamics and interactions that could inspire actionable insights.

By focusing on enabling factors, participants uncover the forces that shape opportunities, creating a comprehensive foundation for modeling futures and understanding their full potential.

Depending upon the size of the Futurecasting session and the amount of data needed, the participants(s) can be tasked to answer all of the experience questions or a subset if time is a factor.

The enabling questions can be broken down into high-level categories to explore all aspects of the EBO approach (Kyle 2008).

For example, participant(s) are tasked to explore potential roadblocks or barriers and think about new business models and practices that would enable the event. Namely, the things that could occur in future that, ultimately, would be advantageous for the organization trying to bring the future about. The important thing is that they look at both sides of the solution or opportunity; for instance, what are the things that will need to be invented or improved over the next 10 years to facilitate the organization's actions and access to bring about the desired future?

Then they also have to contemplate what barriers could be put into the organization's trajectory that could enable them to speed up or collapse the timeline of development, this getting to the solution faster than competitors. Participant(s) also imagine what socio-technical systems and discrete technologies would help facilitate the solution or opportunity and what support systems are required for it to thrive.

Effect-Based Operations at Work in the Futurecasting Method Enabling Questions
In the Futurecasting Method, the concept of effects-based operations helps identify the "center of gravity" for a situation—the one effect or outcome that, if achieved, unlocks the full potential of an opportunity. The goal is to understand the elements that enable the opportunity and how they interact to bring it to life.

By describing the solution or opportunity in detail, participants can pinpoint the event—the physical or digital manifestation of the opportunity—as a key effect that various actors or systems are working to achieve. Enabling questions guide participants to explore how this effect could be supported, amplified, or accelerated.

To break this down, participants can draw inspiration from elements of national power, such as:

- Technological: Innovations or tools that make the opportunity possible.
- Regulatory: Alliances, partnerships, or negotiations that enable the opportunity.

- Information: Communication systems, data flows, or knowledge-sharing mechanisms that support the opportunity.
- Economic: Financial resources, incentives, or markets that drive the opportunity forward.

Beyond these traditional elements, Futurecasting encourages participants to think broadly and creatively about enablers, including:

- Cultural Trends: Shifts in public opinion, popular culture, or values that align with the opportunity.
- Relationships: Networks, collaborations, or social connections that catalyze progress.
- Communication: How messaging, media, or storytelling frames the opportunity and builds momentum.

The key is to consider all aspects—tangible and intangible—that contribute to the realization of the solution or opportunity. The more specific and detailed the exploration, the richer the insights will be for modeling futures.

Putting it into Practice

Enabling questions provide a structured way to map the systems, actors, and conditions that bring an opportunity to life. By answering these questions, participants can build a detailed Effect-Based Model that not only describes the opportunity but also reveals the pathways and interactions that make it achievable. This clarity is crucial for understanding how to harness or enhance these factors to shape a better future.

Systems Thinking and the EBM

Figure 4.5 "... provides a view where whole individuals exist within the center of the helix. In essentially, systems thinking is what occurs when the individual's social, cultural, physical, emotional, and political attributes change the system but at the same time are changed by the collective nature of the system. The lines are an intersection of time. Time infers individuals have levels of experience and familiarity with the system. So, as one comes to know the system, one comes to know the self" (Stalter et al. 2017).

Systems Thinking and EBMs: Unpacking Complexity to Envision Opportunities

Effect-Based Models (EBMs) often encompass a range of experience design and effects-based operations elements, making them rich yet complex tools for Futurecasting. Systems Thinking provides a powerful framework to help analysts better understand and navigate this complexity. It allows the analyst to view the EBM as an interconnected system, where each element influences and interacts with others. By breaking the system into its components, analysts can deeply examine its parts while maintaining a holistic perspective.

Fig. 4.5 Continuum of complexity example showcasing system thinking (Stalter et al. 2017. Courtesy of Wiley)

Nobel laureate and physicist Richard Feynman famously illustrated this approach when he reflected on a glass of wine:

> A poet once said, 'The whole universe is in a glass of wine.' … If we look closely, we see the entire universe. There are the things of physics: the twisting liquid, the reflections in the glass, and the atoms. The glass itself is a distillation of the earth's rocks, revealing the secrets of the universe's age and the evolution of stars. The wine contains arrays of chemicals, ferments, enzymes, and products that speak to the generalization: all life is fermentation… In dividing this glass of wine into parts—physics, biology, geology, astronomy, psychology—remember that nature does not know these divisions. So let us put it all back together and, in the end, enjoy its final purpose: to bring pleasure.

Feynman's reflection embodies Systems Thinking by showing how breaking down a subject into parts—while still understanding their relationships and the greater whole—can uncover deeper meaning.

Applying Systems Thinking to EBMs

In Futurecasting, Systems Thinking allows analysts to approach the EBM as an interconnected network rather than isolated pieces. By identifying how the various parts of the system—technological, social, economic, and cultural—intersect and influence one another, analysts can develop a more nuanced understanding of the opportunity and its broader implications.

A Systems Thinking Example: "Know Thyself"

A practical application of Systems Thinking can be found in healthcare, where researchers created a "helix" map of the human self to improve patient outcomes. This map broke down the complexity of a person into interrelated dimensions:

- Social: The influence of relationships and networks.
- Emotional: The inner world of feelings and mental health.
- Physical: The body and its biological processes.
- Cultural: The shared beliefs and norms shaping behavior.
- Political: The systems of governance and power structures.
- Structural: The frameworks and institutions people navigate.
- Individual: The unique personal identity and choices of each person.

By using a similar approach in Futurecasting, analysts can explore the complexity of opportunities within an EBM, examining not just the individual elements but also their interactions and combined effects.

Why Systems Thinking Matters in Futurecasting

Systems Thinking encourages analysts to take new perspectives, form unexpected connections, and identify patterns within the complexity of an EBM. Whether exploring the development of a sustainable city, the rise of a transformative technology, or the evolution of cultural norms, this approach provides a structured yet flexible way to analyze, understand, and communicate complex opportunities. By deconstructing the EBM into its parts and reassembling it with an awareness of how those parts interact, analysts gain the insights needed to envision richer, more actionable futures.

4.2 Exercise

> All things appear and disappear because of the concurrence of causes and conditions. Nothing ever exists entirely alone; everything is in relation to everything else.
>
> —The Buddha

This exercise builds skills in creativity, imagination, and systems thinking by exploring a person's experience of a future opportunity. By focusing on human-centered narratives and integrating concepts from experience design, practice creating detailed narrative.

Part 1: Build Your Person in a Place Experiencing the Solution or Opportunity

To develop a vivid and relatable narrative, use creativity. By immersing yourself in the person's environment, you'll create a rich narrative that brings the future to life.

Directions

- Describe Your Person: Write a descriptive and creative name for your person.
- List five sensory words to describe your person (e.g., sight, sound, smell, taste, touch).
- Locate Your Person: Identify the neighborhood, city, state/territory/province, and country where your person lives.
- Describe the Scene: Write a few sentences about what's happening. What is the person experiencing? What is the solution or opportunity they're engaging with?
- Enhance the Details: Review your work and add more sensory and descriptive elements to enrich the narrative.

Part 2: Exploring Experience Design Dimensions

This part of the exercise focuses on understanding how a person interacts with a product, service, or experience over time. Pick a product or service you've encountered, and answer the following questions based on key dimensions of experience design.

Dimension 1: Value

- What kind of value does the product provide? Consider functional, financial, emotional, identity, or meaningful value.
- What parts of the product are most valued by people, and why?
- Which value is the strongest priority for users, and how does the product fulfill it?

Dimension 2: Duration

- Initiation: How do people first encounter and begin using the product?
- Immersion: How do people deepen their understanding or interaction with the product over time?
- Conclusion: What happens when people stop using the product? How do they feel about "shutting it down" or setting it aside?
- Continuation: What encourages people to return to the product, and how do they use it again?

Dimension 3: Interaction

- How active or passive is the person's engagement with the product?

- Does the experience require their full attention, or does it integrate seamlessly into their routine?
- How does the product encourage ongoing interaction?

Part 3: Apply Systems Thinking to Enabling an Experience
Using the "Know Thy Self" example of Systems Thinking, analyze the enablers that make an experience possible. This exercise will help you identify and map the systems and interactions that create a future opportunity.

Directions

- Identify the Experience: Write the experience in the center of a page (e.g., ordering groceries online, using an AI-driven personal assistant, or navigating a futuristic urban transit system).
- Map Enablers: Surround the experience with the following categories of enablers:
  - Economic: What business models or financial factors make the experience possible?
  - Technical: What technologies are necessary to enable the experience?
  - Ecosystem: Who are the people, groups, or organizations supporting the experience?
  - Barriers: What obstacles need to be addressed or removed for the experience to occur?
  - Research: What additional knowledge or advancements are needed?
- Analyze Interactions: Explain how these enablers interact and influence each other over time.
- Provide at least two examples of how these interactions shape the experience.
- Reflection.
  - How does integrating sensory details and experience design enrich the narrative of a future opportunity?
  - What insights did you gain by mapping the systems that enable an experience?
  - How might this exercise help you, as an analyst, create more detailed and actionable Effect-Based Models?

By completing this exercise, you'll strengthen your ability to envision opportunities, design human-centered futures, and analyze the systems that bring them to life.

Stories from the Lab
During the pandemic, the Lab adapted to virtual workshops, proving that big ideas can happen anywhere—even from a dining table doubling as an office while wearing PJs. Digital breakout rooms turned out to be a secret weapon, where participants felt freer to share their boldest ideas without the side-eye of a fellow participant. But keeping everyone engaged was a challenge, it often felt like conducting an orchestra where half the musicians accidentally muted themselves. Pre-recorded prompts and asynchronous tasks

became the unsung heroes, keeping the future on track even if the Wi-Fi occasionally wasn't.

An Interview

Brenda Trinidad, Social Scientist, curator of conversations around the socio-cultural dimensions of space exploration and habitation, Analog Astronaut.

Brenda Trinidad, perpetually curious about the world, fascinated in part by the stories we tell each other about our collective life experiences. A few years ago I was introduced as the "Experience Curator" by a colleague during the Las Cruces Space Festival. I loved the way that sounded, how it resonated in my heart. Since then I've adopted that title and have been working to embrace the depths of what that identity means in practice personally and professionally.

As I reflect on my life journey to date, story has always been a part of how I've created meaningful learning experiences. I was trained as a physicist and experimental psychologist and was told stories about what I could and could not do as a woman in sciences. As a higher education professional, I had a chance to help students mitigate the stories heard about what they could and could not do, and together we found ways to find the stories that "fit their truth." When the heart connection was made, I could see it in my students' eyes. I cherish those moments when we celebrated their successes as a result, and I am blessed to remain friends with many of them 10, 20, 30 years later and continue celebrating their successes.

Much of that work was "learning as I go"; it was when I entered graduate school that I found there were people using story in fantastically cool ways that intrigued me. The Human and Social Dimensions of Science and Technology graduate program at Arizona State University is an amazing interdisciplinary program that looks at science and technology through the lens of social sciences and the humanities, such as philosophy, anthropology, and political science. I learned about social justice and speculative fiction as a way to think about the impacts of technological innovation. One of the pivotal classes I took was Creative Nonfiction by Lee Gutkin, the godfather of this genre. He was hired to teach researchers a different way to tell their story of their data. While everyone in the class had data sets from which to use, I was barely starting out. I was the only one that described my "story idea" academically well, but not "a story." I spent the rest of the semester learning how to find "what the story was." The process was a game changer, and I went on to be his teaching assistant to help others during their process of writing their stories. Creative non-fiction as a means to communicate research beyond the walls of academia, along with speculative fiction, to help create futures I want to see? I was hooked.

Participating in the inaugural Emerge Festival in 2012 as part of the research team gave me my first glimpse into speculative fiction in action. (*Produced at ASU from 2012 to 2016 by the author of this book, Cyndi Coon - NOTE added for context related to the book!*) In 2015 I helped design an imaginary Lunar community in line with the Festival theme. I helped design an imaginary Lunar community called Luna City 2175, aimed at creating an immersive simulation for the public to experience "visiting this lunar community." The Festival was held in a theater wherein the stage was transformed into a cross section of our imagined world with actors hired to role play citizens. It was fascinating to see how quickly the public became part of the story as they talked with the actors who successfully stayed in character. I get goosebumps still thinking about it, such an inspiring experience. That Festival was a proof of concept experiment I've yearn to recreate on a more permanent basis. After producing the El Paso Space Festival for a few years, I founded the Southwest Space Institute, LLC. in 2024 to continue putting speculative fiction into action within the space community.

During my graduate research I learned about Analog Astronauts and Analog habitats in which crews of four to six people simulated living and working in a lunar or Martian habitat. Holding on to dreams of being an astronaut, I thought this was the coolest thing ever, and may likely be the closest I would ever get to fulfilling that dream.

In 2022 I got my chance.

In response to an email from the Analog Astronaut community network I applied for an last minute to join a mission in Poland. I was accepted, and off to Poland I went, knowing nothing of the mission, to spend two weeks locked up with four other strangers. The Mission Director asked:

"What role do you want to have?"

"What are my options?"

"How about Communication Officer?

"That sounds cool, yes!"

Mission Support designed and thoughtful experience, but it was the crew that helped make the mission transforming. They were eager, full of energy, hope, and believed that their efforts would make a difference. Non-traditional in nature, our mission was the subject of an award winning film about analog simulations. As a participant and potential curator of such experiences, I was further inspired to find ways to tell the story of simulated missions beyond academic walls.

Now my goal is to continue participating in analog missions to deepen my first-hand knowledge of the simulation experience. In doing so, I will be able to help craft story creation tools for both mission design and debriefing, while shedding light on the importance of this type of experience for the space curious. Ultimately, I want to enhance the crew's ability to think proactively about the impacts of their experience on future celestial communities.

The analog community helps write stories of who can participate in space exploration, and how. These missions missions are no longer restricted to pure scientific research, now including representation from the arts and humanities, and differently abled. Analog missions are becoming increasingly important vehicles to test drive speculative fiction scenarios and stories of as part of actual astronaut training. This is the void my team and

I are working to address through the World's Biggest Analog mission (WBA), scheduled for October 2025.

Through the Southwest Space Institute, LLC. I am the Director of the Speculative Fiction Project for the WBA. One two-week simulated mission, across about seven habitats worldwide, with about 100 crew members. My team is writing speculative fiction to overlay on this mission. This fiction will help tell the story of the mission's potential impact 19 years in future. This is the first time speculative fiction is used as an intentional mission design tool at this level, and we are looking forward to learning what we can during the process. We get to create an imagined future of hope and possibilities, inviting others to help us write this story and find ways to make it real.

We have developed a compelling backstory that will get people excited about the mission. Leaning on much of my past experience described above, we chose to set the story as a future year-long celebration of the pivotal moment the WBA became, setting us on track for thriving lunar and Martian communities in 2044. We picked 2044 specifically because it will be the 50th anniversary of Carl Sagan's "The Pale Blue Dot" book. We have the opportunity to re-introduce a new generation to the impact of Sagan's iconic image and philosophies, along with celebrating additional real and speculative philosophers and scientists. In the process, we also have the opportunity to recontextualize the meaning of February 14th, the day the image was taken in 1992.

(Cyndi) Which is also the anniversary of the Threatcasting Lab.

(Brenda) Exactly. Right.

We will use creative non-fiction tools to tell the story of the actual research and people involved before, during, and after the mission. We plan to produce an anthology of speculative fiction stories highlighting imagined research trajectory between 2025–2044, emulating Project Hieroglyphics produced by the Center for Science and Imagination at ASU.

Our cornerstone message is grounded in the idea of legacy and "being a good ancestor," inspired in part by the book *"The Good Ancestor: A radical prescription for long-term thinking*" by Roman Krznaric. As we develop our story to space curious people as possible to think deeply and broadly about what kind of future we want as an interplanetary species, we continually ask.

"How can you be a good ancestor for future terrestrial and space-based communities?"

Our goal is to encourage people to contemplate being good ancestors through our story as we prepare to expand into the cosmos while improving life on Earth. Our aim with this story, as we tell it from 2044, is to speculate how we succeeded. It is a different way to think about space exploration, a way that can be powerful in creating a shift toward thriving community development terrestrially and in space. As an Experience Curator within the space industry, I am committed to doing what I can to be the good ancestor I hope to see in others.

Chapter Resources

Vocabulary

Phase 2—Futurecasting
The stage in the Futurecasting Method where participants transform insights from Phase 1 into detailed, human-centered visions of possible futures using the Effect-Based Model (EBM).

Effect-Based Model (EBM)
A structured, qualitative model combining research, creative thinking, and participant perspectives to envision specific futures from the standpoint of a person in a place experiencing a problem that is solved or an opportunity realized.

Participatory Design
A collaborative design approach that directly involves participants in co-creating tools, systems, and narratives, ensuring diverse perspectives shape the modeled future.

Experience Design
A method for crafting detailed depictions of how individuals interact with products, services, or opportunities, focusing on sensory, emotional, and contextual elements of the experience.

Science Fiction Prototyping (SFP)
A narrative method for imagining futures through storytelling, integrating research and creative speculation to explore technological, societal, and cultural implications.

Effects-Based Operations (EBO)
An adapted strategic framework from the military, used in Futurecasting to identify enabling conditions, systems, and actions that bring desired futures into reality.

Experience Questions
Guided prompts that help participants describe how a person might encounter, engage with, and be impacted by an opportunity in a future setting.

Enabling Questions
Prompts designed to uncover the systems, actors, and conditions that make a future opportunity possible.

Systems Thinking
An analytical approach for understanding complex, interconnected systems by examining relationships between social, technological, economic, cultural, and environmental elements.

Resources

Primary Inputs for Phase 2

- **Research Synthesis Workbook (RSW)** outputs from Phase 1.
- Curated prompts tied to the research question and time horizon.
- Participant expertise, lived experience, and creativity.
- Reference frameworks for experience design, participatory design, and systems thinking.
- Scenario modeling tools (e.g., spreadsheets, Miro boards, narrative templates).

Supporting Theories & Frameworks

- Science Fiction Prototyping (Johnson 2011).
- Experience Design (Shedroff 2001).
- Participatory Design principles (Robertson and Simonsen 2012).
- Effects-Based Operations (Rickerman 2003).
- Systems Thinking models (Stalter et al. 2017).

Recommendations

1. **Anchor Creativity in Research**—Begin with strong data points from the RSW to keep speculative futures plausible.
2. **Humanize the Future**—Build narratives around a specific person in a specific place to make scenarios relatable.
3. **Embrace Contradictions**—Allow opposing prompts or data points to coexist; the tension often generates richer insights.
4. **Detail the Experience**—Use sensory descriptions and emotional context to deepen engagement with the future scenario.
5. **Think in Systems**—Identify not just what happens, but the enabling factors and interconnections that make it possible.
6. **Balance Micro and Macro Views**—Move fluidly between the personal perspective and the systemic view to create robust EBMs.

7. **Document Everything**—Capture all raw data, even tangential ideas, as they may reveal valuable connections later.

Useful Links

- Multi-Sided Dice link Dice Shop: https://www.thediceshoponline.com/specialist-dice/multi-sided-dice-sets.html
- Applied Future Lab: https://www.appliedfutureslab.com/
- Threatcasting Community: Threatcasting.ai
- Threatcasting Lab at ASU: https://threatcasting.asu.edu/.

References

Glenn J (1972) Futurizing teaching vs. futures courses. Soc Sci Rec 9(3):26–29

Johnson BD (2011) Science fiction prototyping: designing the future with science fiction. Synthesis lectures on computer science. Morgan & Claypool Publishers. https://doi.org/10.2200/S00336ED1V01Y201102CSL003

Karray F, Alemzadeh M, Abou Saleh J, Arab MN (2008) Human-computer interaction: overview on state of the art. Int J Smart Sens Intell Syst 1(1):137–159. https://doi.org/10.21307/ijssis-2017-283

Kyle C (2008) RMA to ONA: The Saga of an effects-based operation [monograph, school of Advanced Military Studies, United States Army Command and General Staff College]. DTIC. https://apps.dtic.mil/sti/pdfs/ADA499725.pdf. https://doi.org/10.21236/ADA499725

Rickerman L (2003) Effects based operations: a new way of thinking and fighting [monograph, School of Advanced Military Studies, United States Army Command and General Staff College]. https://doi.org/10.21236/ADA416050

Robertson T, Simonsen J (2012) Challenges and opportunities in contemporary participatory design. Des Issues 28(3):3–9. https://doi.org/10.1162/DESI_a_00157

Shedroff N (2001) Experience design 1. New Riders Publishing

Stalter AM, Phillips JM, Ruggiero JS, Scardaville DL, Merriam D, Dolansky MA, Goldschmidt KA, Wiggs CM, Winegardner S (2017) A concept analysis of systems thinking. Nurs Forum 52(4):323–330. https://doi.org/10.1111/nuf.12196

Van der Heijden K (2011) Scenarios: the art of strategic conversation, 2nd edn. Wiley

5 Phase 3

You've got to always go back in time if you want to move forward.

—Snoop Dogg

5.1 Backcasting: Time-Phased, Alternative-Action Definition

In Phase 3 of the Futurecasting Method, participants engage in an expanded backcasting process to build on the solution or opportunities envisioned in Phase 2. This exercise helps participants identify actions that can maximize opportunities and recognize the indicators (flags) that signal those opportunities are beginning to manifest. By using a time-phased, alternative-action definition (TAD) process, participants will consider multiple futures with parallel actions, actors, and locations to fully explore how opportunities unfold.

5.1.1 Introduction to TAD Backcasting

Backcasting, first conceptualized by Lovins and Robinson in the 1970s and 1980s, is a technique that begins with imagining a desirable future and works backward to identify the steps needed to achieve it. Unlike traditional strategic planning, which focuses on likely futures starting from the present, backcasting asks: How can we get to the future we want starting there and working backwards to today?

The Futurecasting Method expands on this idea, transforming traditional backcasting into a more dynamic, pluralistic process. Using time-phased, alternative-action definition

B. D. Johnson and C. Coon, *Futurecasting*, Synthesis Lectures on Threatcasting,
https://doi.org/10.1007/978-3-032-21993-0_5

(TAD), participants explore multiple pathways, actors, and actions that can create positive outcomes. This method incorporates parallel timelines and encourages collaboration across diverse sectors, such as academia, government, industry, and nonprofits, to fully realize opportunities in future.

When using Futurecasting's expanded backcasting process to envision a desirable future and map out multiple paths to reach it, participants will identify actions, actors, and indicators (flags) to align efforts and recognize when your envisioned future is beginning to manifest.

These action steps can unfold as follows:

Define the Vision

Start by selecting a desirable future solution or opportunity. This could be a technology breakthrough, a social innovation, or an environmental solution. Use the Effect-Based Model (EBM) developed earlier to ground your vision in specific details. Ask yourself:

- What does this future look like?
- Who benefits from it, and how?
- Why is this future desirable?

Map Time-Phased Actions

Using the TAD approach, work backward from the envisioned future to the present. Break your timeline into two key phases:

0–4 Years: Immediate actions that lay the groundwork for the future.

4–8 Years: Longer-term actions that build on early progress and bring the vision closer to reality.

For each phase, consider:

- Actions: What steps need to be taken (e.g., partnerships, policy changes, investments, technological advancements)?
- Actors: Who needs to be involved (e.g., developers, researchers, community leaders, industries)?
- Locations: Where will these actions occur (e.g., specific cities, industries, or environments)?

Identify Indicators (Flags)

Indicators are the signals that your envisioned future is beginning to take shape. For each phase, define clear and measurable flags that will help track progress. For example:

- A new law or policy is enacted.
- A key technology reaching proof of concept.
- Increased public adoption or funding for the initiative.
- An emerging cultural trend or preference.

Explore Alternative Pathways

Consider multiple parallel futures to ensure flexibility and resilience in your planning. For each alternative pathway, ask:

- What are other ways this future could be achieved?
- How could different sectors (e.g., public and private) contribute in unique ways?
- What redundancies can be built into the plan to account for unexpected challenges?

Visualize Your Backcasting Map

Create a visual representation of your timeline, showing:

- Actions, actors, and locations for each time phase.
- Key flags or indicators.
- Connections between early actions and their second- and third-order effects on later outcomes.

Reflection Questions

- How did breaking the future into phases help clarify the steps required to achieve your vision?
- What unexpected insights emerged from considering alternative pathways?
- How do the flags you identified provide clear signals of progress toward the future you envisioned?

Equip participants with the tools to think systematically and creatively about how to achieve desirable futures. By expanding backcasting into a multi-dimensional process, Futurecasting allows for richer exploration of opportunities and greater adaptability to complex systems (Fig. 5.1).

Fig. 5.1 Visual of traditional backcasting

5.1.2 Review the EBM and the "Experience"

In this phase, participants return to the future scenario and the Experience described in their Effect-Based Model (EBM) from Phase 2. The Experience is the instantiation of the solution or opportunity. The goal is to immerse themselves in the envisioned future, focusing on how to enable and amplify positive opportunities. By leveraging diverse expertise, participants identify actions and preconditions that can shape a better future.

The interdisciplinary nature of the participants is essential for this exercise. Drawing on varied perspectives from academia, government, industry, and communities enriches the exploration of alternative pathways. The detail in the EBM is critical—it provides a rich foundation for participants to fully understand the perspective of the person experiencing the opportunity and the broader systems enabling it.

This deep understanding allows participants to explore how opportunities can unfold, identifying what actions, investments, and collaborations are needed to support their vision of the future.

Key to Backcasting for Opportunities

Begin by reviewing the EBM and the Experience. Put yourself in future world described—what is happening, and what are the defining characteristics of the opportunity? Consider the point of view of both the person experiencing the opportunity and the broader systems supporting it.

Identify Preconditions such as what needs to be in place to enable the opportunity? Consider infrastructure, policies, cultural shifts, technologies, or collaborations that would support the vision. Define actions and sequencing by mapping out the steps required to achieve the future, working backward from the envisioned opportunity to the present. Focus on the sequence of actions by starting with what needs to happen first, and what follows? Leverage diverse perspectives by engaging participants with varied expertise to ensure a holistic exploration.

Encourage input from different sectors to uncover second- and third-order effects.

Reflection Questions

- How did immersing yourself in future help clarify the actions needed to support the opportunity?
- What new insights emerged from identifying preconditions?
- How did diverse perspectives shape your understanding of the sequencing of actions?

By integrating backcasting with the EBM, participants develop actionable pathways to amplify opportunities, ensuring that their Futurecasting insights are grounded in clear, strategic planning.

5.1.3 Gates and Flags

The objective of Phase 3 in the Futurecasting Method is to gather as much detailed information as possible from participants. The facilitator plays a key role in encouraging participants to add specificity and depth to their inputs, which enriches the Effect-Based Model (EBM) and provides a robust foundation for further analysis in the next phase (Fig. 5.2).

Gates: Defining Actions to Enable Opportunities

Gates represent the actions needed to enable, amplify, or optimize a desirable future. They are the critical steps that must be taken by specific actors to make an opportunity possible. Understanding who takes these actions, where they occur, and under what conditions is just as important as defining the actions themselves.

The facilitator should guide participants to move beyond high-level ideas and focus on actionable specifics, identifying key stakeholders and contexts. Gates defined at this stage become the starting points for milestones in the Time-Phased Alternative Action Definition (TAD) process used later.

Example Gates

A global retail company launches an initiative to adopt fully autonomous delivery vehicles, requiring partnerships with tech firms, government approvals for regulations, and large-scale investments in infrastructure.

- R&D Innovator—Technological Breakthroughs: A pharmaceutical company funds targeted research in synthetic biology to develop customizable treatments for rare diseases, collaborating with academia and venture capitalists to accelerate progress.
- Government Regulator—Policy and Standards: A national government creates data privacy laws for autonomous AI systems, requiring businesses to comply with stringent security standards before deployment.
- Industry Association—Collaborative Standards: A consortium of cloud providers forms an industry-wide coalition to establish interoperable standards for data storage, driving competition and reducing costs for smaller enterprises.
- Academia—Future Workforce Development: A university develops a specialized MBA program focused on sustainability leadership, training executives to drive environmental, social, and governance (ESG) strategies in corporations.

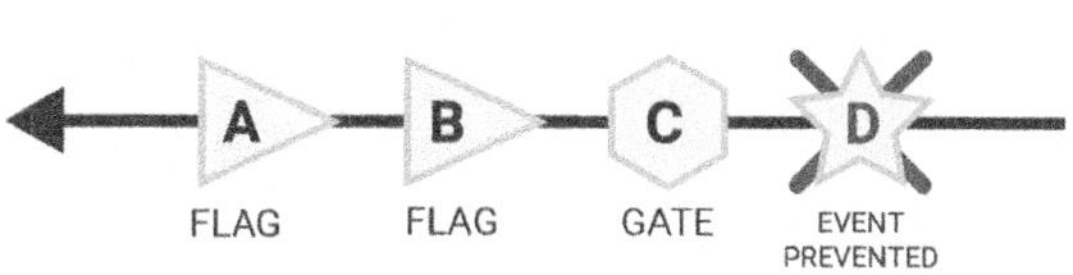

Fig. 5.2 Example of futurecasting workbook focused on gates

Facilitators should prompt participants to focus on:

- Specific Actors: Who performs the action? Is it an individual, a team, or an organization? Examples might include private-sector innovators, public policymakers, or NGOs.
- Context: Where and how will the actions take place? Are they region-specific, industry-wide, or global in scope?
- Measurable Outcomes: What are the tangible results or milestones? This could range from increased market share to the successful launch of a product or policy.

By collecting this level of detail, participants generate actionable data that forms the backbone of the EBM and ensures that subsequent phases of Futurecasting are built on well-defined, practical insights.

Flags
See Fig. 5.3.

Flags: Indicators of Emerging Futures
Flags are significant events or developments—technological, scientific, economic, cultural, or geopolitical—that signal the progression of an envisioned future. While stakeholders may not control these events, their occurrence creates irreversible path dependencies with lasting consequences. Flags serve as early indicators, or "signals," that opportunities (or challenges) are materializing.

To qualify as a flag, an event must create a permanent change in the trajectory of the envisioned future. For instance, a technological breakthrough, a major policy shift, or a cultural revolution could irreversibly alter the landscape, making it impossible to return to the prior state.

Characteristics of Flags

- Irreversible Impact: Once a flag occurs, it creates lasting change and establishes dependencies that define future pathways.
- Iterative Nature: Flags often build sequentially, each enabling the next. For example, Flag A may lead to Flag B, which in turn sets the stage for Flag C. This progression ensures that flags align logically and meaningfully with the envisioned future.
- Wide-Ranging Sources: Flags can stem from advancements in technology, demographic changes, cultural shifts, or external disruptions such as economic recessions or natural disasters.

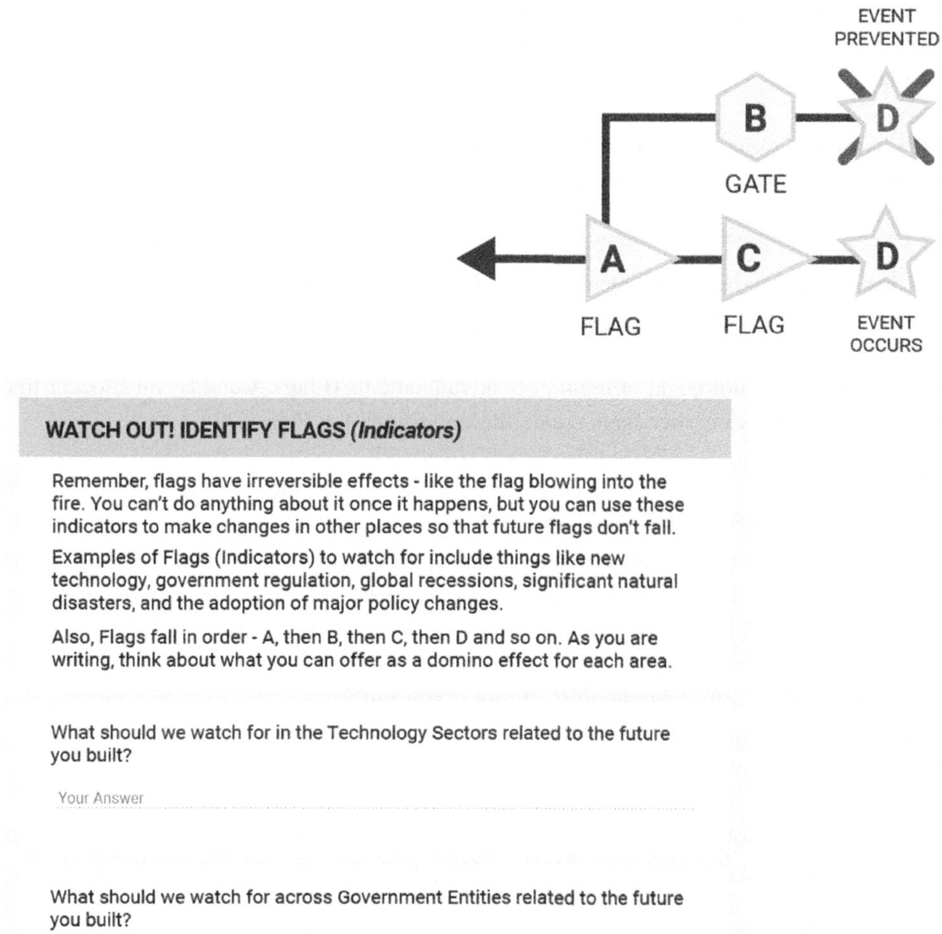

WATCH OUT! IDENTIFY FLAGS *(Indicators)*

Remember, flags have irreversible effects - like the flag blowing into the fire. You can't do anything about it once it happens, but you can use these indicators to make changes in other places so that future flags don't fall.

Examples of Flags (Indicators) to watch for include things like new technology, government regulation, global recessions, significant natural disasters, and the adoption of major policy changes.

Also, Flags fall in order - A, then B, then C, then D and so on. As you are writing, think about what you can offer as a domino effect for each area.

What should we watch for in the Technology Sectors related to the future you built?

Your Answer

What should we watch for across Government Entities related to the future you built?

Fig. 5.3 Example of futurecasting workbook focused on flags

Examples of Flags

- A groundbreaking AI technology becomes commercially viable, changing industry norms.
- A country enacts a sweeping climate regulation that shifts global supply chains.
- The adoption of a novel healthcare policy reshapes access to emerging therapies.
- A global economic downturn accelerates automation in certain sectors.
- An environmental disaster catalyzes new infrastructure standards in vulnerable regions.

During workshops, participants should identify key flags that signal the progression of the envisioned future. Facilitators can guide participants to add depth and specificity by asking:

- What needs to happen before this flag can occur?
- What changes in technology, policy, or cultural norms might enable this flag?
- What would follow this flag, and how does it connect to subsequent events?

For instance, if participants identify Flag C (e.g., widespread adoption of autonomous vehicles), prompt them to consider:

- What early technological or regulatory advancements (Flags A and B) must occur first?
- How would these earlier flags create the conditions for Flag C?

This process ensures that flags are logically sequenced, interconnected, and meaningful within the context of the envisioned future.

Flags are invaluable for monitoring progress and preparing for change. By identifying and tracking early indicators, businesses and organizations gain a clearer understanding of when to act and how to allocate resources effectively. Well-defined flags:

- Provide early warning signs of emerging opportunities;
- Help organizations avoid premature actions or overreactions;
- Enable timely, informed responses when subsequent gates (actions) are triggered.

For example, a global company anticipating widespread adoption of renewable energy might track the development of critical enabling technologies (Flag A), followed by government subsidies (Flag B), and eventually mass-market adoption (Flag C). These flags ensure that the company invests at the right moment, avoiding unnecessary risk while maximizing impact.

Reflection Questions

- How do the flags you identified create a logical and meaningful progression toward the envisioned future?
- What early signals or events could act as precursors to the flags?
- How can tracking these flags inform strategic decisions in your organization or industry?

By articulating detailed flags and their sequential connections, participants ensure that the envisioned future is actionable, measurable, and grounded in plausible developments. Flags become not just indicators of what's coming but tools for shaping how to respond effectively.

The Duality of Gates and Flags

In the Futurecasting Method, gates and flags often exist in a dual relationship. One organization's gate—an actionable step within their control—might simultaneously serve as a flag, or signal, for another entity. This perspective shift underscores the interconnected nature of the data generated in Phase 3.

Gates and flags represent the highest level of detail in backcasting. Gates are actions you or your organization can take to enable, amplify, or accelerate an opportunity. Flags, on the other hand, are external events or conditions beyond your control but which indicate that an opportunity is beginning to materialize.

Why Diverse Perspectives Matter

Achieving specificity in identifying gates and flags often requires domain expertise. A broad mix of perspectives ensures that blind spots are minimized, and possibilities are expanded. For example:

When Futurecasting the future of farming, including a farmer provides insight into agricultural processes, while a technologist can highlight advancements in precision farming or AI tools.

If analyzing opportunities in cybersecurity, it's essential to include not just technologists but also legal professionals and human resources representatives who might bring unexpected considerations to light. Diverse backgrounds enrich the conversation, ensuring that gates and flags are robust, actionable, and relevant across multiple domains. Homogeneous groups, by contrast, may struggle to think beyond their shared experience.

Flags: Indicators to Watch for

Flags serve as early indicators of opportunities. Once identified, they act as signposts signaling when it's time to take action. For example, during the backcasting process, participants define flags that suggest an opportunity is gaining momentum. When a flag is triggered, it prompts the organization to consult its playbook and initiate the appropriate gate.

Flags are essential for strategic preparation because they allow organizations to act at the right moment, neither too early nor too late.

Gates: Actions to Enable the Future

Gates are the actionable steps that an organization can take to seize opportunities. These actions might require significant investment or change, especially when planning for the long term. The 10-year timeframe used in Futurecasting is particularly valuable here—it provides the runway needed to implement transformative changes, such as overhauling a technical infrastructure or developing entirely new business models.

Gates are more effective when designed with clear specificity and a phased approach. By laying out a timeline of steps, organizations can align their resources and efforts to create meaningful progress toward their envisioned future.

Creating a Playbook

The interplay between gates and flags forms the basis of an actionable playbook. The backcasting process identifies:

Flags: External events or indicators to monitor.
Gates: Specific actions to take when those indicators arise.

For example, in cybersecurity:

A flag might be a new government regulation mandating stricter data privacy measures.
The corresponding gate could be implementing advanced encryption protocols to ensure compliance.
By incorporating insights from diverse perspectives, the playbook becomes a dynamic resource that spans multiple domains, ensuring readiness to respond to opportunities as they arise.

The Value of the 10-Year Timeline

The decade-long horizon allows organizations to plan strategically and take meaningful action. Major initiatives—such as adopting new technologies, building infrastructure, or reskilling a workforce—require time to execute effectively. The 10-year window ensures that organizations can break down these initiatives into manageable steps, aligning them with gates and flags to stay on track.

Reflection

- How can the duality of gates and flags help align actions across different organizations or teams?
- What specific domains or perspectives could enrich your identification of gates and flags?
- How might a well-crafted playbook support your organization's ability to adapt and seize opportunities over a decade-long timeframe?

By leveraging the dual nature of gates and flags, Futurecasting provides a structured yet flexible approach to navigating and shaping complex futures.

Incrementalism
What Is Incrementalism?

Incrementalism is a decision-making approach many people use instinctively, even without naming it. It's the process of taking small, logical steps to achieve a goal. For example, getting dressed involves incremental steps that follow a logical order—like putting on socks before shoes.

When applied to organizations, incrementalism takes on a strategic dimension. Charles E. Lindblom, the political scientist who introduced the concept in the 1950s, described it as "a *method of building upon past actions to shape future decisions*." Rather than seeking sweeping, value-maximizing solutions, organizations often rely on smaller, more manageable steps to navigate uncertainty and complexity.

Why Use Incrementalism?

Incrementalism is effective because it aligns with real-world constraints and psychological realities. Its benefits include:

- Practicality: Small steps are easier to implement, track, and adjust.
- Responsiveness: It adapts well to uncertainty, allowing flexibility as conditions change.
- Complexity Management: Breaking big problems into smaller parts makes them more approachable.
- Shared Plans: Incremental actions are easier to communicate and gain consensus on, especially in diverse teams.

Incrementalism in Action

Incrementalism works particularly well in environments where complexity, uncertainty, or rapid change are present. In Futurecasting, for instance, incrementalism allows practitioners to implement gates (specific actions) and monitor flags (indicators) in manageable phases, ensuring the organization progresses toward its long-term vision without being overwhelmed.

By taking incremental steps, organizations can align past experiences with present conditions to craft practical, forward-looking strategies. As Lindblom noted, this approach is less about striving for perfection and more about making consistent, logical progress.

5.1.4 Milestones

See Fig. 5.4.

Milestones: Building Actionable Timelines with TAD Backcasting

Milestones are critical markers in the development of a Futurecasting timeline. They represent significant points of action and progress, expanding on the gates identified earlier. By mapping these milestones onto a timeline, participants can visualize the iterative nature of actions and their early impacts, ensuring a well-structured path toward the desired future.

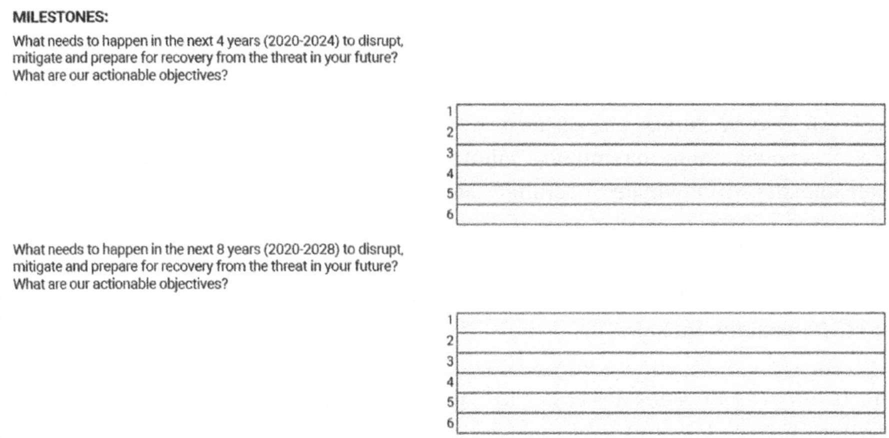
MILESTONES:

What needs to happen in the next 4 years (2020-2024) to disrupt, mitigate and prepare for recovery from the threat in your future? What are our actionable objectives?

1
2
3
4
5
6

What needs to happen in the next 8 years (2020-2028) to disrupt, mitigate and prepare for recovery from the threat in your future? What are our actionable objectives?

1
2
3
4
5
6

Fig. 5.4 Example of futurecasting workbook focused on milestones

Using the Time-Phased Alternative Action Definition (TAD) backcasting approach, milestones become more detailed, including specifics such as physical locations and the people or organizations responsible for each action. This clarity ensures that participants can anticipate the steps needed to enable future actions, offering time to plan, monitor, and adjust.

Breaking Down the 10-Year Timeline

There's no universal rule for dividing the 10-year span into actionable segments, but the default framework uses four-year and eight-year milestones:

Eight Years Out: Critical actions must occur by this point to prevent delays or missed opportunities.
Four Years Out: Foundational actions are taken to enable the eight-year milestones.
Immediate Actions: Planning and initial steps begin now to meet four-year goals.

For example, if a law or regulation needs to change by year eight, the organization must start influencing policymakers, lobbying, or building advocacy networks well before year four.

The TAD Backcasting Process

TAD backcasting generates a complex matrix of interconnected actions, locations, and actors. Participants often work non-linearly, moving back and forth along the timeline to identify enabling actions and intermediate steps.

Key considerations include:

- Interdependence: Each milestone must logically build upon the previous one.
- Actors: Who is responsible for each milestone (e.g., government, private industry, or NGOs).
- Locations: Where will these actions take place (e.g., specific regions, industries, or communities)?

Iterative Process

Completing the TAD backcasting is a collaborative and iterative process. It's common for participants, especially those new to Futurecasting, to refine their milestones over multiple rounds. Facilitating two to three iterations of Phases 2 and 3 can help participants generate a richer volume of data, uncover new insights, and refine the envisioned future.

Practical Advice on Milestones

Dr. Jon Brickey, a Futurecasting practitioner, provides insight into using milestones effectively:

> Milestones are essentially gates but with much more detail laid out over time. For example, if you're trying to change a regulation, that doesn't happen overnight. You'll need intermediate steps—hiring a lobbying firm, building internal advocacy, creating a network of stakeholders—all of which need to happen years before the law changes.
>
> The process also reveals decision points and multiple pathways. You don't need all the answers upfront. By backcasting, you uncover key steps and can adapt to new possibilities. It's structured brainstorming, enabling preparation for multiple futures.

Reflection

- How do your milestones build logically toward the envisioned future?
- Are there areas where additional specificity or expertise is needed?
- How can the timeline help your organization prepare for multiple possible futures?

This structured approach ensures participants create a roadmap that is flexible, actionable, and aligned with the complexities of the Futurecasting process. Milestones provide the clarity needed to move confidently toward a desirable future while allowing space for adaptation and innovation.

5.1.5 Second- and Third-Order Effects

A significant advantage of the Futurecasting Method is its ability to uncover second- and third-order effects of future actions. This is particularly valuable for large, complex organizations such as global corporations or defense agencies. These cascading impacts often reveal opportunities and challenges that might otherwise remain hidden.

The use of Science Fiction Prototypes (SFPs) makes these visions tangible. SFPs allow organizations to visualize and immerse themselves in potential futures, fostering clarity and dialogue around necessary actions.

"People aren't wired to imagine the future, ten or even five years out, which is a blocker to innovation," says Kate O'Keeffe, Senior Director of Cisco's Hyper-Innovation Living Labs (CHILL), who applied Futurecasting in 2017 to explore digital supply chain opportunities. "We need to create that world for them, so they can immerse themselves in this future scenario, making it immediately apparent what kind of solutions we need to prepare for that future."

Sharing the Vision: Reporting Out

At the end of Phase 3, participants share their work by presenting the SFPs they've created. They tell the story of their envisioned future, highlighting key opportunities and describing the timeline of actions that enable that future.

This reporting-out session is an integral part of the process. It allows the broader group to hear the detailed scenarios developed in smaller groups, sparking new ideas and cross-pollination of perspectives.

Iterative Exploration for Richer Futures

Futurecasting Phases 2 and 3 are typically repeated three to four times with the same group of participants. Each iteration enables participants to refine their approach, gain comfort with the materials, and delve deeper into the possibility space.

New Foundational Data Points: For each round, participants select fresh data points from the research prompts, ensuring a diversity of explored futures.

Expanding the Possibility Space: Moderators encourage participants to broaden the scope of their SFPs, exploring more detailed futures and identifying additional second- and third-order effects.

Broader Organizational Insights

As participants cycle through the phases, their comfort level with the methodology grows, leaning to richer, more varied insights. The iterative process reveals a wider array of future opportunities and actionable paths. The cumulative effect is a set of robust, imaginative, and actionable future scenarios that provide organizations with a clearer understanding of what steps to take.

Reflection

The reporting-out sessions and iterative rounds of Futurecasting create a dynamic environment where participants build upon each other's ideas. The process generates actionable insights while fostering a culture of innovation and future readiness.

Chapter Resources

Vocabulary

Phase 3—Backcasting (TAD)
The stage in the Futurecasting Method where participants work backward from a desirable future to identify actions, actors, locations, and indicators that enable the envisioned opportunity, using the Time-Phased, Alternative-Action Definition (TAD) process.

Time-Phased, Alternative-Action Definition (TAD)
An expanded backcasting approach that maps multiple pathways, parallel timelines, and diverse actors to achieve a future vision, broken into immediate, mid-term, and long-term phases.

Gates
Specific, actionable steps within an organization's control that enable or accelerate the realization of a desirable future.

Flags
External events or indicators, often outside an organization's control, that signal a future opportunity is beginning to materialize.

Milestones
Key markers on the TAD timeline that represent significant achievements or progress points, often corresponding to critical gates.

Preconditions
Systems, resources, policies, or cultural conditions that must be in place before a gate can be executed.

Alternative Pathways
Multiple, parallel routes to achieving a desired future, often involving different sectors, actors, or strategies to ensure resilience and adaptability.

Second- and Third-Order Effects

Cascading consequences—intended and unintended—that emerge from a particular action or milestone, influencing other parts of the system.

Incrementalism

A decision-making approach that breaks big goals into small, manageable steps, allowing adaptation to uncertainty and complexity over time.

Resources

Primary Inputs for Phase 3

- Completed Effect-Based Models (EBMs) and "Experience" descriptions from Phase 2.
- Participant expertise spanning multiple sectors (public, private, nonprofit, academia, communities).
- Time horizon planning tools (10-year framework with 0–4 year, 4–8 year, and 8–10-year segments).
- TAD mapping templates for actions, actors, locations, and indicators. Sector-specific data and historical precedent for policy, technology, and cultural shifts.

Reference Frameworks & Tools

- Traditional backcasting (Lovins, Robinson);
- TAD methodology for multi-path, parallel timeline mapping;
- Effects-based operations (EBO) for identifying enabling factors;
- Systems thinking for interdependencies and cascading effects;
- Science fiction prototyping for illustrating second- and third-order impacts.

Recommendations

1. **Start with the Future in Sharp Detail**—Revisit the EBM and fully immerse in the envisioned opportunity before mapping actions backward.
2. **Map Preconditions Before Actions**—Identify infrastructure, policy, cultural, and resource requirements before sequencing gates.
3. **Separate Gates and Flags Clearly**—Ensure actions within control (gates) are distinguished from external indicators (flags) for better readiness planning.
4. **Build in Alternative Pathways**—Map multiple routes to the same outcome to maintain flexibility in uncertain environments.

5. **Think in Milestones, Not Just Steps**—Mark significant achievement points that help measure progress along the 10-year arc.
6. **Layer in Incrementalism**—Break big shifts into smaller, feasible actions to reduce resistance and manage complexity.
7. **Consider Second- and Third-Order Effects**—Explicitly explore ripple effects, both positive and negative, that could reshape the pathway.
8. **Use Diverse Perspectives for Richer Gates & Flags**—Include domain experts and unconventional thinkers to expand the possibility space.
9. **Make It Visual**—Create TAD maps that clearly show timelines, actors, locations, and cause-effect linkages for easier communication and tracking.
10. **Iterate for Depth**—Repeat Phases 2 and 3 multiple times to refine and expand insights.

Additional Key Tools

- **Gate and Flag Worksheets** for detailing specific, measurable actions and identifying external signals to monitor.

5.2 Exercise

Backcasting is traditionally visualized as a single path that connects a future scenario to the present. The Futurecasting Method, however, expands this process into a time-phased, multi-path exploration that is proactive and opportunity-focused. Using the TAD (Time-phased, Alternative-action Definition) approach, participants create a rich map of gates (actions) and flags (indicators) to enable desirable futures. This exercise will guide you through mapping opportunities using parallel timelines, multiple actors, and diverse actions.

Objective
To apply the TAD backcasting approach to identify opportunities, explore multiple pathways, and create actionable timelines with clear indicators for progress.

Steps

Step 1: Define the Opportunity
Begin by identifying a desirable future opportunity. This could be a technological advancement, a social innovation, or a policy outcome. Use the following questions to clarify the opportunity:

- What is the desired outcome?
- Who benefits, and why is it important?
- Where and when does this opportunity take place?

Step 2: Create a Traditional Backcast
Visualize a single timeline connecting the future opportunity to the present. Use gates and flags to represent:

- Gates (Actions): Steps that enable the opportunity (e.g., passing legislation, launching a new technology, or creating public awareness).
- Flags (Indicators): Signs that the opportunity is beginning to manifest (e.g., pilot programs, market shifts, or cultural changes).
- Draw this single-path timeline to establish a baseline understanding.

Step 3: Explore Branching Futures
Expand the backcast by exploring how the opportunity could evolve differently across locations, actors, or actions. Create branching paths to represent variations in how the future could manifest. For example:

- Geographic Branching: Different countries or regions adopting distinct policies or strategies.
- Actor Branching: Different groups (e.g., governments, industries, communities) taking separate actions toward the same opportunity.
- Action Branching: Parallel efforts occurring simultaneously (e.g., investments in complementary technologies or grassroots movements in tandem with policy changes).
- Illustrate these branches, noting how different gates and flags emerge along each path.

Step 4: Incorporate Parallel Timelines
In some cases, timelines may progress at different rates or with varying outcomes. For example:

- A policy change might be enacted quickly in one country but delayed in another.
- A technology might be adopted more rapidly by one industry while another lags behind.
- Identify these parallel paths and note how they interact, influence each other, or diverge entirely.

Step 5: Map Complex Systems with TAD Backcasting
Using the branching and parallel paths, build a comprehensive map of your opportunity's future. Include:

- Multiple gates and flags distributed across timelines, actors, and actions.
- Connections between early actions and their second- and third-order effects.

Reflection

- How did branching and parallel timelines enhance your understanding of the opportunity?
- What unexpected insights emerged about the actions or actors needed to enable the future?
- How might the results of your backcasting map be used to inform decisions, policies, or strategies?
- How will your backcasting map guide decision-makers in shaping this opportunity?
- What actions can be prioritized to create momentum toward the future you've envisioned?

By completing this exercise, you'll develop a detailed and flexible roadmap that highlights how to enable desirable futures across multiple contexts, ensuring actionable insights for the Futurecasting process.

Stories from the Lab
Flags and gates are the backbone of Futurecasting, but explaining them to new comers can be a comedy of errors. During one workshop, a participant earnestly asked, "Wait, so are the flags literal? Like, are we planting them somewhere? And what kind of gate are we talking about? A garden one?"

Cue the facilitator's hasty analogy: "No, the flags are like warning signs—like seeing your gas light come on before a road trip. And gates? They're more like checkpoints—what you need to pass to keep the journey going."

Another participant chimed in: "So, flags are the red alerts, and gates are the boss battles?"

From then on, every time someone hit a key insight, someone else would yell, "Level up! You've unlocked a gate!" Who knew Futurecasting could double as a game night?

An Interview
Dr. Claire A. Nelson, Forbes 50 Female Futurist, Sustainability Engineer & Innovation Consultant, Keynote Speaker & Storyteller, White House Champion of Change

I am a futurist, a storyteller, social entrepreneur, and a sustainability engineer. I got into futures thinking because I worked for many years in international development. Those projects tend to be 30 and 40 year projects. And as an engineer, I felt that our design was less than adequate or less than theologically effective because we were using historical data to design for the future. At the time when I started thinking about this, it was not really accepted practice in the international financial institutions to think about foresight, and nobody really took me on. But I did begin to apply strategic foresight in some of my work, especially, in terms of having the conversations with societies that we're trying to change. So you're trying to do a social project around healthcare or around agriculture and you are asking communities of interest, stakeholders to believe in this project, and they're not seeing what you are seeing because they're buried in the day to day, their living lives that are very hard, impoverished lives. And so very quickly I realized that my storytelling vocation and abilities were going to be of high use to me. And lo and behold, there was something called scenarios. So scenarios and storytelling have a lot in common. Scenarios which is one tool in future thinking and strategic foresight is used to explore the landscape of one possibility in the various possibilities that the futures could merge into. And I realized that I'm a perfect match for scenarios because of my storytelling knack. And so from early days, I started using storytelling and training people on storytelling.

And then one day I found out about the Emerge Festival of the Future at Arizona State University, and I thought I'd like to create a story about the future. And so I pitched and was selected to do a future of sports, and I use the opportunity to craft a story from my personal future, which I now call Moon Runnings. The life and times of the first Jamaican on the moon there began my, um, furry into what I call future telling or future steal or narrative foresight, because I used the opportunity to really build out a world, um, in which I imagined a life and people doing space tourism on the moon. And so I had to construct an imaginary world. So it wasn't quite science fiction because I was using ideas which I think are realistic and have a higher probability of existing in the near future because it was a near future story. I was talking about 2035 using existing dates, which are already talked about in NASA. I wrote and I performed my story, and I had a fantastic time doing so. And at the time, my purpose for doing this was really to have a conversation about how do we achieve the UN Global SDGs and writing this about the year 2035 and beyond for the Emerge Festival of the Future of Sport was set for 2040. So I was also projecting beyond 20, 30, 10 years after, what were some of the issues on the table at the time in terms of the global narrative, the global narrative for humanity.

I am a global futurist. I always focused on big global concepts such as energy systems, water systems, food security, and human security and I do this through casting future stories.

Chapter Resources

Vocabulary

Phase 3—Backcasting (TAD)
The Futurecasting stage where participants work backward from a desirable future to map the actions, actors, locations, and indicators needed to enable it, using the Time-Phased, Alternative-Action Definition process.

Time-Phased, Alternative-Action Definition (TAD)
An expanded backcasting approach that explores multiple parallel pathways, actors, and timelines for achieving a future vision, broken into short-, mid-, and long-term phases.

Gates
Specific, controllable actions that enable, amplify, or accelerate the realization of a desirable future.

Flags
External events or conditions, often outside direct control, that signal a desired future is beginning to emerge.

Milestones
Key achievements on the TAD timeline marking measurable progress toward the envisioned future, often tied to gates.

Preconditions
Foundational systems, policies, resources, or cultural conditions required before a gate can be executed.

Alternative Pathways
Multiple, parallel routes toward a desired outcome, designed to ensure resilience and adaptability under different future conditions.

Second- and Third-Order Effects
Cascading, often unexpected consequences—positive or negative—resulting from a particular action, decision, or milestone.

Incrementalism
A decision-making approach that advances toward a goal in small, logical steps, allowing for adaptation in uncertain or complex environments.

Resources

Primary Inputs for Phase 3

- Completed Effect-Based Models (EBMs) and "Experience" narratives from Phase 2.
- Participant expertise spanning multiple domains (public, private, nonprofit, academia, communities).
- Research data on technological, cultural, economic, and policy trends.
- 10-year planning framework with 0–4 year, 4–8 year, and 8–10 year phases.
- TAD mapping templates for actions, actors, locations, and indicators.

Reference Frameworks & Tools

- Traditional backcasting (Lovins, Robinson);
- TAD multi-path mapping methodology;
- Systems thinking approaches for mapping interdependencies;
- Effects-based operations (EBO) concepts for identifying enabling factors;
- Science fiction prototyping for illustrating future worlds and ripple effects;
- Gate & flag worksheets for structured data capture.

Recommendations

1. **Immerse in Future First**—Begin by revisiting the EBM and "Experience" to ground backcasting in a detailed, lived view of the opportunity.
2. **Map Preconditions Before Sequencing Actions**—Identify necessary infrastructure, cultural shifts, and resources before placing gates on the timeline.
3. **Separate Gates from Flags Clearly**—Keep controllable actions (gates) distinct from external signals (flags) to ensure readiness and accurate tracking.
4. **Use Multiple Pathways for Resilience**—Avoid overreliance on a single route; map alternative strategies across sectors and actors.
5. **Think in Milestones**—Define achievement points along the 10-year arc to measure progress and maintain momentum.
6. **Apply Incrementalism for Complex Change**—Break large-scale transformation into smaller, achievable steps to manage risk and build buy-in.
7. **Identify Second- and Third-Order Effects Early**—Anticipate ripple impacts to spot hidden opportunities or risks.
8. **Engage Diverse Perspectives**—Include unconventional thinkers and cross-domain experts to enrich gate and flag identification.

9. **Make it Visual**—Use diagrams, branching timelines, and cause-effect maps to communicate complexity clearly.
10. **Iterate for Depth**—Repeat Phases 2 and 3 multiple times, refining gates, flags, and pathways for richer insights.

Useful Links

- Multi-Sided Dice link Dice Shop: https://www.thediceshoponline.com/specialist-dice/multi-sided-dice-sets.html
- Applied Future Lab: https://www.appliedfutureslab.com/
- Threatcasting Community: Threatcasting.ai
- Threatcasting Lab at ASU: https://threatcasting.asu.edu/.

Phase 4

6

Your mind is for having ideas, not holding them.

—David Allen.

6.1 Post-analysis, Synthesis, and Findings

The Futurecasting Method culminates in a critical phase: post-analysis. After completing the workshop and gathering raw data, this phase focuses on synthesizing the information to develop actionable findings. These findings illuminate possible and potential futures, identify specific actions (gates), and establish indicators (flags) to guide strategic decision-making (Fig. 6.1).

Purpose of Post-analysis

The post-analysis phase transforms raw workshop outputs into actionable insights. By organizing, reviewing, and synthesizing data, analysts create a roadmap of opportunities, offering organizations clear pathways to enable desirable futures. The final findings help teams prioritize actions and monitor key indicators as the envisioned future unfolds.

Key Activities in Post-analysis

- Prepare the raw data for analysis.
- Organize all data collected during the workshop, including workbooks, participant feedback, and outputs from iterative cycles.
- Digitize and centralize the data for ease of access and systematic review.

B. D. Johnson and C. Coon, *Futurecasting*, Synthesis Lectures on Threatcasting,
https://doi.org/10.1007/978-3-032-21993-0_6

POST-ANALYSIS
TIMELINE

RESEARCH QUESTION:

Team/Round	#1 "Summary"	Notes	#2 "Meaning"	Notes	#3 "Novelty"	Notes
	(For Part 1/2/3 of Workshop Workbook)		(For Part 1/2/3 of Workshop Workbook)		(For Part 1/2/3 of Workshop Workbook)	
T1/R1						
T2/R1						
T3/R1						
T4/R1						
T1/R2						
T2/R2						
T3/R2						
T4/R2						

Fig. 6.1 Post-analysis timeline

- Perform multiple rounds of analysis.
- Conduct iterative reviews to refine insights and uncover patterns in the data.
- Group similar themes, identify emergent opportunities, and validate findings against the foundational research prompts.
- Generate, validate, and review findings.
- Use structured frameworks to develop findings that connect the workshop's outputs to actionable futures.
- Collaborate with diverse stakeholders to ensure findings are robust and reflective of multiple perspectives.
- Specify actions (gates) and indicators (flags).
- Define the specific actions necessary to enable opportunities or mitigate potential risks.
- Identify indicators that signal when actions should be initiated, ensuring readiness for emerging trends and events.

Deliverables of Post-analysis

The outcomes of this phase are synthesized into an action report which serves as a tool for sharing insights with key stakeholders. This report includes:

- Findings: Detailed descriptions of possible and potential futures.
- Actions (gates): Practical steps to enable opportunities.
- Indicators (flags): Early warning signs or signals to monitor progress toward the envisioned future.

Reflection Questions

- How can the findings from this phase influence organizational strategy?
- What processes ensure the data is synthesized into clear, actionable insights?
- How can the defined gates and flags help teams stay proactive and adaptive?

This phase of the Futurecasting Method ensures that the creative outputs of the workshop are systematically refined into a strategic guide, enabling organizations to act decisively and confidently in shaping their desired futures.

6.1.1 The Role and Art of the Analyst(s): Turning Data into Futures

With the workshop complete and a wealth of raw data in hand, it's time to move into phase 4: post-analysis. This phase is where analysts synthesize the collected information into actionable findings, crafting a small set of possible and potential futures, actionable insights, and key indicators. From this rich dataset, analysts will distill a focused selection of five futures to form the foundation of narratives and a final action report.

The role of the analyst is essential to this process. The unique expertise, experience, and perspective bring invaluable depth and nuance to the analysis. Just as the workshop participants were chosen for their diversity, the analyst(s) contributions are equally vital to ensure a balanced and innovative exploration of the data.

What Analyst(s) Do in Phase 4

- Review and Process Raw Data—Analyst(s) sift through the data collected in the workshop, using the Futurecasting framework to identify patterns, insights, and connections. Depending on the scope and scale of the exercise, this may involve extensive collaboration among multiple analysts.
- Leverage Individual Expertise and Perspectives—Each analyst brings their own lens—shaped by their background, biases, and relationship to the research question. These distinct viewpoints are critical for uncovering new opportunities and enriching the findings.
- Collaborate and Debate—When multiple analysts are involved, their collaboration and constructive debates allow for deeper validation of ideas. Diverse perspectives often reveal unexpected insights and new dimensions within the raw data.
- Diversity Among Analysts: A Key Strength—Just as every pianist has a unique way of playing the same instrument, each analyst approaches the Futurecasting framework with their own style. While the methodology provides structure and consistency, the artistry lies in how each analyst interprets and applies it.

To maximize the effectiveness of the post-analysis phase, assembling a diverse team of analysts can be a strategic advantage. For instance:

- Domain Expertise: If Futurecasting opportunities in banking and connected technologies, having analysts with expertise in financial systems, cybersecurity, or fintech innovation will provide depth.
- Broader Perspectives: Including analysts with experience in underserved communities, generational consumer behaviors, or emerging cultural trends can uncover opportunities that more traditional perspectives might overlook.
- Interdisciplinary Collaboration: When analysts from different fields collaborate, they can identify novel patterns, connect dots across domains, and generate richer, more holistic findings.

Understanding Analyst Contributions

It's important for analysts to recognize and articulate their own expertise, experience, and biases. These factors shape how they engage with the data and the research question, influencing the insights they contribute. By intentionally curating a team with complementary skills and perspectives, organizations can ensure a robust analysis process.

When synthesizing raw data into findings, analysts should:

- Contextualize Data: Relate raw insights to the research question, ensuring alignment with the Futurecasting goals.
- Challenge Assumptions: Encourage debate within the team to refine and validate interpretations.
- Iterate: Revisit findings multiple times to capture the fullest range of insights and opportunities.

Reflection

- How does your expertise influence the way you interpret the raw data?
- What new perspectives could you bring to the findings through collaboration?
- Are there gaps in the analysis where additional perspectives could add value?

By blending structure and creativity, the analysts' role in phase 4 ensures the raw data becomes a powerful tool for envisioning and enabling future opportunities.

6.1.2 Prepare the Raw Data Sets for Analysis

The transition from workshop outputs to actionable insights begins with post-analysis preparation, where raw data is cleansed, structured, and organized for seamless processing. This crucial step ensures that the data is both accurate and usable for analysts while safeguarding participant anonymity.

Data Hygiene

Data hygiene refers to the process of cleaning and standardizing the raw data to optimize its usability. This ensures consistency across the dataset, minimizes errors, and prepares it for analysis. Key actions include:

- Standardize Data—Ensure all data conforms to a uniform format (e.g., proper placement in cells, consistent labeling, and structured entries).
- Validate Data—Check for missing or incomplete data, ensuring no critical information is overlooked. Analysts may decide whether to follow up with participants or leave gaps if appropriate.
- De-duplicate Entries—Identify and remove redundant or repeated sections of data to avoid skewing results.
- Analyze Data Quality—Review the dataset to confirm it contains enough depth and detail for meaningful analysis.
- Investigate and Repair Issues—Resolve inconsistencies or errors, ensuring the dataset is ready for the next phase.

Structured data ensures the analyst(s) can work efficiently, whether using manual methods or automated tools. This preparation lays the foundation for robust analysis.

Anonymization

While participant identities may remain linked to their contributions during the analysis phase for clarity and follow-ups, the final dataset and subsequent reports should be anonymized. This step ensures participants can freely share ideas without concern for attribution and protection of personal information as all raw data excludes personally identifiable information (PII), such as names or affiliations, safeguarding participant privacy. By removing PII, analysts also comply with ethical standards, ensuring that findings can be shared broadly without compromising individual identities.

Summarizing the Raw Data

Summarizing raw data is a critical step to condense and clarify the findings while preserving the essence of the scenarios described. A summary provides:

- A concise overview of key points.
- Separation of components (e.g., person, event, implications) for clarity.
- Steps for summarization.
- Review of the raw data for speed-read for an initial understanding, noting key elements such as actors, events, and implications.
- Clarity of Understanding—Reread carefully, ensuring comprehension of all details and the participant's intent.
- Notes—Use bullet points to capture concise, essential information. Avoid extraneous details or overly complex language.

Why Summarization Matters

Summarizing reduces the raw data from sprawling, multi-page workbooks to manageable, digestible insights. This process immerses the analyst in the scenario while providing clarity for subsequent analysis rounds.

Reflection Questions

- Is the dataset consistent and free from errors or redundancies?
- Does the summary maintain the original intent and tone of the raw data?
- Have all participant identities been fully anonymized for ethical reporting?

By combining rigorous data hygiene, thoughtful anonymization, and effective summarization, analysts ensure the raw data is ready to be transformed into meaningful findings during the next phases of the futurecasting method.

6.1.3 Perform Multiple Rounds of Analysis on the Raw Data

The post-analysis phase of the Futurecasting Method is an evolving practice, with analysts increasingly exploring a blend of cognitive and digital tools to enhance their work. While this chapter focuses on the core tasks and outcomes of each step in the process, the choice of tools and procedures remains flexible to suit the needs of the research question, team, and analyst preferences (Fig. 6.2).

Three Rounds of Analysis

The post-analysis phase typically involves three rounds of analysis to extract key findings from the raw data. Each round progressively refines and organizes the data to uncover insights and actionable opportunities. Selecting the right tools and methods for these rounds is part of the art and skill of the analyst.

POST-ANALYSIS
TIMELINE

Round	Team	Round 1 "Summary"
1	Red Pawn	Lorem ipsum dolor sit amet, consectetuer adipiscing elit, sed diam nonummy nibh euismod tincidunt ut laoreet dolore magna aliquam erat volutpat. Ut wisi enim ad minim veniam, quis nostrud exerci tation ullamcorper suscipit lobortis nisl ut aliquip ex ea commodo consequat. Duis autem vel eum iriure dolor in hendrerit in vulputate velit esse molestie consequat, vel illum dolore eu feugiat nulla facilisis at vero eros et accumsan et iusto odio dignissim qui blandit praesent luptatum zzril delenit augue duis dolore te feugait nulla facilisi. Lorem ipsum dolor sit amet, cons ectetuer adipiscing elit, sed diam nonummy nibh euismod tincidunt ut laoreet dolore magna aliquam erat volutpat. Ut wisi enim ad minim veniam, quis nostrud exerci tation ullamcorper
1	Orange Pawn	Lorem ipsum dolor sit amet, consectetuer adipiscing elit, sed diam nonummy nibh euismod tincidunt ut laoreet dolore magna aliquam erat volutpat. Ut wisi enim ad minim veniam, quis nostrud exerci tation ullamcorper suscipit lobortis nisl ut aliquip ex ea commodo consequat. Duis autem vel eum iriure dolor in hendrerit in vulputate velit esse molestie consequat, vel illum dolore eu feugiat nulla facilisis at vero eros et accumsan et iusto odio dignissim qui blandit praesent luptatum zzril delenit augue duis dolore te feugait nulla facilisi. Lorem ipsum dolor sit amet, cons ectetuer adipiscing elit, sed diam nonummy nibh euismod tincidunt ut laoreet dolore magna aliquam erat volutpat. Ut wisi enim ad minim veniam, quis nostrud exerci tation ullamcorper
1	Yellow Pawn	Lorem ipsum dolor sit amet, consectetuer adipiscing elit, sed diam nonummy nibh euismod tincidunt ut laoreet dolore magna aliquam erat volutpat. Ut wisi enim ad minim veniam, quis nostrud exerci tation ullamcorper suscipit lobortis nisl ut aliquip ex ea commodo consequat. Duis autem vel eum iriure dolor in hendrerit in vulputate velit esse molestie consequat, vel illum dolore eu feugiat nulla facilisis at vero eros et accumsan et iusto odio dignissim qui blandit praesent luptatum zzril delenit augue duis dolore te feugait nulla facilisi. Lorem ipsum dolor sit amet, cons ectetuer adipiscing elit, sed diam nonummy nibh euismod tincidunt ut laoreet dolore magna aliquam erat volutpat. Ut wisi enim ad minim veniam, quis nostrud exerci tation ullamcorper
1	Teal Pawn	Lorem ipsum dolor sit amet, consectetuer adipiscing elit, sed diam nonummy nibh euismod tincidunt ut laoreet dolore magna aliquam erat volutpat. Ut wisi enim ad minim veniam, quis nostrud exerci tation ullamcorper suscipit lobortis nisl ut aliquip ex ea commodo consequat. Duis autem vel eum iriure dolor in hendrerit in vulputate velit esse molestie consequat, vel illum dolore eu feugiat nulla facilisis at vero eros et accumsan et iusto odio dignissim qui blandit praesent luptatum zzril delenit augue duis dolore te feugait nulla facilisi. Lorem ipsum dolor sit amet, cons ectetuer adipiscing elit, sed diam nonummy nibh euismod tincidunt ut laoreet dolore magna aliquam erat volutpat. Ut wisi enim ad minim veniam, quis nostrud exerci tation ullamcorper

Fig. 6.2 Example of post-analysis workbook after round 1

The Post-analysis Workbook

A dedicated post-analysis workbook is used to document every step of the process. This serves multiple purposes:

- Collaboration: Provides a shared space for multiple analysts to work cohesively.
- Transparency: Captures all steps, ensuring accountability and clarity in the process.
- Efficiency: Acts as a reference when drafting the final report.

Rounds 1 and 2: Axial Coding

The initial rounds of analysis employ axial coding, a technique widely used in qualitative research.

Definition of Axial Coding:

Axial coding involves identifying relationships between categories and their subcategories. It seeks to:

- Pinpoint central phenomena or themes (the "axis") in the data.
- Reassemble or disaggregate data to highlight connections and distinctions within categories.

As explained by Anselm Strauss and Juliet Corbin in Basics of Qualitative Research:

> The essence of axial coding is to identify some central characteristic or phenomenon (the axis) around which differences in properties or dimensions exist. Axial coding is therefore a process of reassembling or disaggregating data in a way that draws attention to the relationships between and within categories (Wicks 2010).

Using Axial Coding in Futurecasting

In the context of Futurecasting, axial coding helps analysts:

- Identify key themes emerging from workshop data.
- Relate these themes to specific opportunities, actions, or trends.
- Create a structured narrative that connects participant insights to the broader research question.

Flexibility in Tools
While axial coding is a foundational technique, analysts are encouraged to explore additional methods and tools. Whether using digital platforms for data visualization, AI-driven text analysis, or traditional manual techniques, the goal is to find the most effective way to synthesize the data into actionable findings.

Round 1: Summary
The first step in post-analysis is round 1, where the analyst(s) work to summarize and organize the raw data collected during the Futurecasting workshop. The goal is to distill each envisioned future into concise, manageable summaries, providing a clear overview of the participant(s)' imagined futures. At this stage, analysts focus on capturing the essence of the models without interpretation, ensuring fidelity to the original intent of the participants.

Once round 1 is complete, the data becomes significantly more accessible, allowing the analyst(s) to focus on identifying specific actions (gates) and indicators (flags) in subsequent phases.

Key to Round 1 Length Reduction

The first objective is to reduce the extensive raw data from multiple pages into a concise text block. Typically, these summaries are one to two paragraphs, capturing the critical components of the Futurecasting EBM (Parts 1–3).

What to Include: Focus on the most relevant aspects related to the research question.

What to Exclude: Leave out excessive detail unless a specific element is crucial to the scenario.

Summarization involves crafting a brief but accurate overview of the original material while maintaining its core meaning and intent.

What Is a Summary?

A summary provides a high-level view of the material, emphasizing key facts while omitting unnecessary details. This process ensures clarity and accessibility for subsequent rounds of analysis.

Steps for Effective Summarization

- Get a general idea of the data.
- Speed-read the material to gain an initial understanding of its content. Focus on headings, key details, and any recurring themes.
- Check your understanding.
- Reread carefully to ensure your interpretation aligns with the participant(s)' intent.
- Take notes.
- Use bullet points to capture the most important ideas. Start each bullet with a keyword or phrase and include only one point per bullet. Keep notes concise and well-organized.

Tip: Use frameworks like the Cornell Note-Taking System to structure your notes for easy reference.

Write the Summary

Draft a cohesive paragraph that condenses the data into its essential points. Refer to your notes rather than the original text to ensure clarity and simplicity. Verify that the summary accurately reflects the original content and includes all critical points. Remove unnecessary details and refine the language for clarity.

Formatting the Summaries

To make the data more digestible, organize the summaries under clear headings:

- Opportunities Derived from Prompts: Highlight key possibilities envisioned by the participant(s).
- Person: Provide a brief description of the individual at the center of the scenario, including their context and environment.

- Event: Summarize the pivotal event and its implications on the individual and their broader community.
- Benefits of Round 1 Summarization.
- Accessibility: Condensed summaries make it easier to reference and analyze the data in subsequent rounds.
- Immersion: Analysts become deeply familiar with the participant(s)' scenarios, aiding in later analysis.
- Clarity: Summaries provide a clear narrative structure, ensuring alignment with the research question.
- Challenges and Considerations.

While summarization is invaluable, some details may be smoothed out or lost. Analysts may need to revisit the raw data to confirm specific elements or clarify ambiguities. Striking the right balance between brevity and completeness is crucial.

Reflection

- Does the summary accurately reflect the participant(s)' vision?
- Are all essential elements included without overloading the text?
- Is the summary structured to align with the goals of the Futurecasting process?

By systematically reducing and summarizing the data in Round 1, analysts lay the groundwork for deeper exploration and more focused insights in the later phases of post-analysis.

Round 2: Meaning

Using the output from Round 1, the analyst(s) is tasked to review each summarized future and capture the meaning of the future in context to the Futurecasting Foundation (topic, research question, and application areas), the prompts and any additional information from discussions that took place during generation of the raw data.

The analyst(s) can explore the following questions:

- How does the summary address the topic?
- How does the summary answer the research question?
- Does the future provide a new perspective on the foundation?
- What is the significance of the future to the research foundation?
- What is essential and what can be left out?

In round 2, the analyst(s) begin to interpret the data. Analyst(s) look for interesting themes within each scenario or that span across multiple scenarios. Analyst(s)' measure of "interesting" is somewhat arbitrary, but generally it fills one of two criteria: either they are unique to a scenario and stand out as an outlier, or they link with several other scenarios and could lead to a finding.

Round 2 will greatly reduce the detail from Round 1, identifying clusters and patterns from the raw data. During round 2, the summary is reduced to its high level meaning with shorthand notes for possible clusters or patterns (see cross round techniques discussed later in this chapter).

There are two primary techniques the analyst(s) use during round 2:

- Broader connotation.
- Prompts comparison.

Broader Connotation

Using the summaries from round 1, the analyst(s) compares the futures against the research foundation to understand what connotation they might have. In other words, what does the summary of the future mean to the topic, research question, and application areas?

The analyst(s) can explore the following questions:

- Does it uncover new implications?
- Does the future have a broader meaning to the foundation beyond the future itself?
- Are there any topic areas that stand out as special?
- Are there broader actions at work?

Prompts Comparison

As the name indicates, analyst(s) will compare the summaries of the futures to the information in the prompts. Considering all of the prompts, the analyst(s) can explore the following questions:

- What topics, areas, or issues does the future address?
- Is there a single theme that stands out?
- Is there a contradiction to the prompts?
- Is there a unique detail that was surprising?

When thinking about how to record the meaning in the EBMs, there are several techniques to code the ideas identified in the data into your post-analysis workbook. The first is to use the words of the scenario itself to describe the effects of the model. This is called in vivo coding. Another way to think about these patterns is to describe them in gerunds (nouns turned into "-ing" verbs). Generally, short phrases and bullet statements are usually appropriate for finding meaning, because the emphasis is on simplifying complex ideas into simple building blocks ("divergent thinking") before recombining them into larger clusters later ("convergent thinking").

Round 3: Novelty

Round 3 is the final round in the post analysis. Using the output from rounds 1 and 2, the analyst(s) explores what is novel, or what stands out from the previous rounds. This novelty should directly answer or give a new perspective to the research foundation.

This is where the art of the analyst(s) is applied.

- What is the data telling you?
- What was meant?
- Are new issues uncovered in context to the research foundation?
- How is the research question answered in a way that has not been addressed previously?
- What futures are new in the range of the possible and probable?
- What information might be useful to take specific actions in the application areas?
- Are there any clusters, patterns, or outliers?

Transforming these lists of themes (from round 2) into distinctive categories that illustrate the "novelty" of a particular future or group of futures comprises round 3. Rapidly moving from the individual themes in round 2 to novel findings in round 3 requires both disciplined thinking and experiential points of view to answer the "so what?" of the matter.

A novel finding can be a new concept to the organization or as a new instance of something seen elsewhere. Another way to approach novelty is a concept that will have profound implications to the application area. This kind of novelty is usually unprecedented and may not correlate or transfer to other situations.

Across all three rounds there are techniques available to the analyst(s) to help in the identification of novelty. Consider these to be cross-round techniques.

- Clustering.
- Reflection.
- Outliers.

Clustering

During each round the analyst(s) looks for clusters and patterns in the data. Clusters and patterns can take on nearly any form including terms, words, themes, futures, characters, technologies, details, etc. The analyst(s) looks for any commonality in recurring actors, technologies, places, etc. that might appear in multiple EBMs.

Throughout the post analyst, the analyst(s) notes these clusters, calling them out in the analyst workbook.

Outliers

In every round of post-analysis, particularly round 3, analyst(s) are tasked with identifying and evaluating outliers—those unique, unexpected, or contradictory elements that stand apart from the broader dataset. These outliers often hold valuable insights and can illuminate opportunities or perspectives that might otherwise be overlooked.

Key Questions to Guide Outlier Analysis—What Stands Out?

- Is there a specific opportunity, actor, or technology that diverges from the broader trends?
- Does this outlier provide a fresh perspective or redefine the Futurecasting Foundation?
- Does the outlier challenge or conflict with other envisioned futures?
- How might this tension reveal new insights or areas for further exploration?

Unspoken Narratives: What Was Not Discussed?

Analysts should also pay attention to gaps or blind spots in the raw data. By examining what wasn't said, new areas for exploration may emerge. Guiding Questions:

- Were certain topics, groups, or technologies noticeably absent from the discussions?
- Examples:
- A specific demographic or community.
- A critical or emerging technology.
- A geographic region with unique challenges or opportunities.
- Does the data reveal any biases that may have shaped participant responses?

Opportunities for Further Exploration

Outliers and omissions are not weaknesses; they are opportunities to broaden the scope of exploration. Use these insights to:

- Propose new prompts for future workshops.
- Develop additional research questions.
- Highlight potential areas of bias or oversight to refine the Futurecasting Foundation.

Reflection

Outliers often represent the unexpected sparks that push Futurecasting into uncharted territory. By carefully examining what stands out—and what is left unsaid—analysts can uncover deeper opportunities and ensure a more holistic view of possible futures.

6.1.4 Generate, Validate, and Review Findings

The clusters and patterns identified during the three rounds of analysis form the foundation of the findings. These findings directly address the research question, providing fresh perspectives and actionable insights that enhance the entire Futurecasting Foundation.

What Are Findings?

Findings can:

- Highlight specific possible and potential opportunities for innovation or advancement.
- Define broader opportunity areas that warrant further exploration.
- Serve as a bridge between the raw data and the actionable steps outlined in the Futurecasting process.

Validating Findings

Validation ensures that findings are credible, robust, and reflective of the data collected. Analysts have several methods to verify their findings, fostering collaboration, refinement, and accuracy.

Collaborative Validation

When working as a team, analysts may conduct the three rounds of analysis independently before convening to compare results.

Collaboration allows:

- Refinement of findings through discussion.
- Exposure of new insights that might have been identified by only one analyst.
- Identification of potential biases (implicit or explicit) in individual analyses.

Subject Matter Expert (SME) Review

SMEs whose research or work served as prompts in the Futurecasting workshop can provide valuable validation. Their review ensures the findings are grounded in relevant expertise and align with the latest developments in their fields.

Participant Peer Review

Analysts can share their findings with the original workshop participants for validation.

This approach ensures:

- Perspectives and opinions from participants are accurately captured.
- An opportunity for feedback and clarification of any misinterpretations.

Why Validation Matters

Validation is not just about confirming what was found but also about refining and expanding insights. Collaborative and peer-reviewed findings are stronger, richer, and more reliable, offering a comprehensive view of possible and potential futures. By leveraging diverse inputs and perspectives in the validation process, the findings become a more robust tool for decision-making, shaping innovative pathways, and addressing the research question.

6.1.5 Specify Actions to be Taken and Indicators to be Monitored

Once the findings have been thoroughly validated, the next step is to translate them into actionable strategies and measurable indicators. This step is critical for bridging the insights from Futurecasting into real-world applications, ensuring that opportunities identified in the findings are both actionable and trackable over time.

Actions (Gates)

Actions, or gates, represent specific, tangible steps that can be taken to capitalize on opportunities or address emerging futures. Analyst(s) revisit the raw data, post-analysis workbook, and backcasting results to identify and refine these actions.

Key Steps for Defining Actions:

- Review clusters and EBMs.
- Examine how the EBMs have been clustered, combined, or nested within the findings.
- Identify high-priority gates across these clusters.

Map Actions to Findings

Ensure that every action is directly tied to a finding, making it clear how the step advances or supports the opportunity identified.

Leverage Additional Resources

Refer to the prompts, conduct further research, or engage with SMEs and participant(s) to refine or expand the identified actions.

Examples of Gates:

- Corporate Innovation: Launching an R&D initiative focused on emerging technologies highlighted in the findings.

- Policy Development: Drafting regulations to encourage equitable access to identified opportunities.
- Community Engagement: Partnering with local stakeholders to co-create solutions.

Indicators (Flags)

Indicators, or flags, act as signposts that the envisioned future is beginning to materialize. These provide early warning or confirmation that specific trends, opportunities, or challenges are developing in alignment with the findings.

Key Steps for Identifying Indicators:

- Examine backcasting.
- Review the backcasting for each EBM to identify milestones and events that align with findings.
- Link indicators to actions.

Each indicator should directly relate to a corresponding action, helping to track its effectiveness or the progression of the identified opportunity.

Expand Through Additional Input

Use prompts, research, or SME interviews to uncover potential indicators not previously identified.

Examples of Flags:

- The adoption rate of a new technology in specific markets.
- Legislative changes or the introduction of new industry standards.
- Significant cultural or demographic shifts related to the findings.
- Connecting actions and indicators to the application area.

To ensure relevance, the analyst(s) revisit the application areas outlined in the Futurecasting Foundation. They expand on the findings, actions, and indicators to create a comprehensive roadmap for implementation.

This final step ensures that the outputs of Futurecasting are actionable, measurable, and aligned with the intended goals of the project. By doing so, the process not only envisions possible and potential futures but also lays the groundwork for realizing those opportunities in meaningful ways.

Exercise

There is no exercise for this section. Instead, refer to the Project in Part 2, where you'll find detailed guidance on running a Futurecasting exercise tailored to your needs. Dive in there to apply the concepts hands-on and start envisioning your own future.

Stories from the Lab

During one workshop, a participant hit on an idea so out-of-left-field that the room fell silent. After a moment, someone said, "Wait, are we allowed to do that?" The facilitator grinned and replied, "That's the point. If you feel like you're breaking the rules, you're probably doing it right."

From that moment on, participants competed to one-up each other with bizarre but brilliant ideas. One team proposed autonomous city pigeons that deliver snacks (and only occasionally insults), while another suggested algae farms in skyscraper lobbies. By the end, the novelty wasn't just in the ideas—it was in how much fun everyone had leaning into the weirdness. Turns out, it isn't just about the future; it's also about rediscovering the joy of asking, "What if?"

An Interview
Dr. Divya Chander

CEO, Founder, Futurist Neuroscientist, Physician

I use future forecasting or Futurecasting because it lets me fully inhabit the present while also co-inhabiting a parallel world where things are as they should be. In that place, there is no friction, no needless struggle, just a kind of shimmering alignment. I can see the path from here to there as if it were a tunnel or wormhole, a liminal bridge connecting now to what could be. Sometimes I feel like I'm the only one who sees it, and I want to turn to everyone and say, "Do you see this? Do you see how extraordinary this could be?" That moment of recognition fuels my motivation to manifest it, to make the unseen tangible.

I live and work in a landscape of relentless technological change, where breathtaking possibilities coexist with profound risks. Futurecasting is how I imagine taming dual-use technologies, putting in place the strategies that align them with the good while protecting people and systems from vulnerabilities and threats. I often perceive parallel futures, parallel versions of myself inhabiting them, and that awareness drives me to choose the ones that are inclusive, equitable, and designed to lift everyone, especially the most vulnerable.

For me, Futurecasting is not about passively accepting the trajectory we're on—it's about deliberately creating the futures we want to inhabit. It's exhilarating to envision what that could look like, to step into it in my mind, and then to work toward co-creating it with others. But it's never about unchecked creation. The world is not a zero-sum game, and imagining the future is an act of care as much as invention. I want to protect as much as I want to innovate, to nurture as much as I want to transform. Futurecasting allows me to hold both of those impulses at once: to dream boldly and to guard wisely, so that what we build does the most good.

Chapter Resources

Vocabulary

Phase 4—Post-analysis
The phase in the Futurecasting Method where raw workshop data is organized, synthesized, and validated to produce actionable findings, gates, and flags.

Findings
Synthesized insights from workshop data that describe possible and potential futures, highlight opportunities, and guide decision-making.

Gates
Specific, actionable steps organizations can take to enable, amplify, or accelerate an opportunity.

Flags
Indicators or signals—often external and outside direct control—that a future opportunity is beginning to materialize.

Post-analysis Workbook
The structured document used to track the three rounds of analysis, coding, clustering, and validation.

Axial Coding
A qualitative analysis technique that identifies relationships between themes and categories, used in rounds 1 and 2 to connect data points.

In Vivo Coding
Coding directly with the words or phrases used by participants in their scenarios, preserving original intent and tone.

Round 1—Summary
The first stage of post-analysis, condensing raw scenario data into concise summaries while maintaining fidelity to participant intent.

Round 2—Meaning
The second stage, interpreting summaries to extract thematic meaning and link scenarios to research questions and prompts.

Round 3—Novelty
The final stage, identifying what is new, unique, or significantly impactful in the data, including outliers and gaps.

Clustering
Grouping related ideas, actions, or themes across multiple scenarios to identify patterns.

Outliers
Unique or unexpected elements in the data that stand apart from patterns but may reveal valuable opportunities or insights.

Anonymization
Removing personally identifiable information (PII) from datasets to protect participant privacy.

Resources

Primary Inputs for Phase 4

- Raw participant data from all Futurecasting workshop phases.
- Research prompts and foundational documents.
- Effect-based models (EBMs) from phase 2 and backcasting maps from Phase 3.
- Analyst expertise, both domain-specific and cross-disciplinary.
- Tools for qualitative data analysis (manual or digital).

Supporting Frameworks and Methods

- Axial coding (Strauss & Corbin).
- In vivo coding for participant-authored language.
- Clustering techniques for theme identification.
- SME review for technical accuracy.
- Participant peer review for fidelity to intent.

Recommendations

1. **Prioritize Data Hygiene**—Standardize, de-duplicate, and validate all datasets before analysis begins.
2. **Respect Participant Privacy**—Apply anonymization early to protect identities and foster trust.
3. **Maintain Fidelity to Original Intent**—Summaries should preserve tone, meaning, and participant perspective before interpretation.

4. **Leverage Multiple Analyst Perspectives**—Collaboration across disciplines strengthens findings and reduces blind spots.
5. **Use Iterative Rounds for Depth**—Progress from summarization to interpretation to novelty for richer insights.
6. **Document the Process**—Keep a transparent post-analysis workbook to track decisions and analysis steps.
7. **Treat Outliers as Assets**—Unique elements may reveal overlooked opportunities or gaps in the research.
8. **Link Every Action and Indicator to a Finding**—This ensures recommendations are directly grounded in the analysis.
9. **Validate Findings Rigorously**—Use SMEs, peer review, and cross-team comparison before finalizing.

   Keep the End-User in Mind—Findings should be clear, actionable, and relevant to the application area.

Useful Links:

- Multi-Sided Dice link Dice Shop: https://www.thediceshoponline.com/specialist-dice/multi-sided-dice-sets.html
- Applied Future Lab: https://www.appliedfutureslab.com/
- Threatcasting Community: Threatcasting.ai
- Threatcasting Lab at ASU: https://threatcasting.asu.edu/

Reference

Wicks D (2010) Coding: axial coding. In: Mills AJ, Durepos G, Wiebe E (eds) Encyclopedia of case study research, vol 1. SAGE Publications, Thousand Oaks, CA, pp 153–155

Phase 5

7

> You can be the best futurist, analyst and modeler on earth. You can come up with exact pictures of possible and potential futures BUT if you can't communicate it appropriately to multiple audiences so that they can understand it and use it to take action … then you have failed.
>
> —the authors

7.1 Output

In Chapter seven we explore the best-known methods for capturing and presenting findings from the Futurecasting Method. The goal is to equip analyst(s) with strategies to effectively communicate their insights to diverse audiences. Since individuals and organizations have unique preferences for receiving information, selecting the appropriate format and delivery method is crucial for ensuring the findings resonate and drive action.

7.1.1 How to Deliver Findings

Since Futurecasting is an applied Futures Method, the practical application of its findings is essential to the final phase. As analyst(s) craft their outputs, they must always consider two key questions:

- Who will use these findings?
- How will these findings be used?

B. D. Johnson and C. Coon, *Futurecasting*, Synthesis Lectures on Threatcasting,
https://doi.org/10.1007/978-3-032-21993-0_7

The ultimate value of Futurecasting lies in how well the findings are communicated and whether they empower others to take action or shape strategy. Packaging findings into clear, compelling narratives ensures that they resonate with the intended audience and provide actionable insights that inspire meaningful progress.

Key Considerations for Delivery

- Audience-Centric Focus: Tailor the delivery to the needs, expertise, and expectations of the target audience.
- Actionable Narratives: Frame findings in a way that not only explains possibilities but also motivates and equips users to act.
- Strategic Relevance: Ensure that the findings align with the organization's or individual's goals, enabling them to integrate insights into their strategy effectively.

By keeping the end users and their needs in mind, analyst(s) can transform raw data into tools for innovation, decision-making, and long-term planning.

7.1.2 Traditional Methods for Delivering Findings

See Fig. 7.1.

The Technical Action Report

A technical action report is a foundational way to present the findings of a Futurecasting Workshop. This comprehensive document includes a detailed overview of the Futurecasting Method, the Futurecasting Foundation (topic and research question), and the results and findings from the analyst(s)' post-analysis. It also provides access to raw data for transparency and further review.

A key strength of the technical report is its transparency. By clearly presenting the inputs, research, and outputs, readers can quickly grasp the reasoning behind the findings and proposed actions, making it easier for organizations or individuals to adopt and act upon them.

While not subject to formal academic or peer review, technical reports often undergo internal reviews with workshop participants and subject matter experts (SMEs). This feedback loop ensures accuracy, invites additional insights, and refines the findings before publication.

Sample Outline/Table of Contents

- Executive fummary.
- Overview of Futurecasting.
- Introduction.

Fig. 7.1 .

- Futurecasting workshop goals.
- Opportunities identified.
- Flags (indicators).
- Gates (actions).
- Conclusion.
- Appendices with raw data.

To make the report more engaging, consider including short vignettes throughout the text. These brief, story-like summaries offer a glimpse into the imagined futures, much like movie trailers, making the findings more relatable and memorable.

Technical reports are typically distributed as polished text documents (e.g., PDFs) and may include visual elements like charts or illustrations to enhance accessibility.

Academic Papers

Publishing findings in academic journals offers an opportunity to reach specialized audiences, such as researchers, professors, or multidisciplinary professionals. While a technical report provides a broad overview, an academic paper hones in on one or two key findings, offering deeper analysis for scholarly or professional communities.

When writing an academic paper, start by choosing who your target audience is. Ask yourself who needs to know about your findings. Are they specialists in one domain or multidisciplinary?

Next, select a research journal relevant to your topic and audience. Understand their content guidelines and review recent publications to ensure alignment. Your academic paper can take the form of research articles, technical analyses, or position pieces, tailored to the chosen journal's style.

Brown (2017) emphasizes that publishing in academic journals has three main benefits: enhancing credibility, establishing practitioner expertise, and creating a lasting, searchable record.

Executive Summaries, Briefings, and Presentation Decks

Not all audiences have the time or need for a lengthy report. For those who need a concise overview, executive summaries, briefings, or presentations are ideal formats to deliver highlights of the findings.

Key formats for delivering outputs include concise executive summaries that distill the most critical insights into one to two pages, briefings designed to fit anywhere from 10 to 60 min depending on the audience's interest and needs, and presentation decks that provide a streamlined visual narrative of the findings, typically limited to nine or ten slides for clarity and impact.

Example Presentation Structure

Slides 1–3: Introduce the opportunity and provide contextual research using language tailored to the organization.

Slide 4: Establish your credibility as the analyst.

Slides 5–8: Present key findings, opportunities, indicators, and recommended actions.

Slide 9: Conclude with a call to action, such as requesting feedback, offering the full report, or proposing the next steps.

A well-crafted summary or presentation piques interest, providing just enough information to encourage deeper engagement with the full report. The ultimate success of Futurecasting findings depends on their application. By carefully tailoring the format and narrative to the audience's needs, analyst(s) can ensure that findings are not only understood but also acted upon, driving innovation and strategy across various domains.

7.1.3 Alternative Outputs to Deliver Findings

While traditional outputs like technical reports, academic papers, and executive summaries are effective, they don't always capture the attention of senior leaders or decision-makers who are pressed for time. To address this, alternative formats can convey Futurecasting findings in an engaging and digestible way, sparking imagination and making the concepts relatable. These outputs are especially useful for illustrating human, ethical, political, or organizational opportunities while highlighting the second- and third-order effects of these futures.

Graphic Novels and Comic Books

Graphic novels and comic books provide a highly visual and engaging way to share Futurecasting findings. They transform speculative futures and scenario narratives into compelling, accessible stories. Using illustrations inspired by Science Fiction Prototypes (SFPs) and vignettes, these formats allow readers to immerse themselves in the imagined future while grasping its broader implications.

The strength of this medium lies in its ability to evoke emotion and imagination. For instance, a graphic novella could explore a future where artificial intelligence and robotics are so advanced that microscopic "bugs" conduct surveillance. By visualizing this, the reader can better understand potential opportunities and ethical dilemmas, such as privacy concerns or technological misuse.

Adding an introduction or afterword can ground the story in the analyst(s)' findings, linking the narrative to the broader Futurecasting research.

Podcasts, interviews, and roundtable discussions offer another dynamic way to present findings. These formats allow analyst(s) to convey complex ideas conversationally, making them relatable and easy to digest. The goal is to capture the audience's attention by highlighting key ideas and sparking interest. A podcast episode could focus on the most compelling opportunities from the findings, weaving in real-world analogies or expert commentary. This format works well for audiences who might prefer listening during their commute or multitasking over reading long reports. Podcasts can also be a gateway to deeper engagement, providing links to detailed reports or supplementary materials.

The Impact of Alternative Outputs

These alternative formats are not just for creative flair; they are strategic tools to ensure that Futurecasting findings resonate with diverse audiences. By tailoring the medium to the audience's preferences, analyst(s) can increase the reach and influence of their work, empowering individuals and organizations to act on the insights. Whether through a captivating graphic novel or an insightful podcast, the ultimate goal is to make Futurecasting findings actionable and memorable.

Output Resources

An Interview

Margaret Orr, owner and graphic designer MorrDesign

Why use a graphic designer to design a report—Why that's important?

The reason that graphic designers exist is not to make things pretty. It's really about strategically getting your message across. It's about strategic communication because if the message is not received, then the work is pointless.

So what a graphic designer can do is make sure that your target audience receives the message you're trying to convey. And that uses things like imagery or layout, using white space and creating a place for the eye to rest or hierarchy in where a title this big and then a subtitle is a little smaller, and then there's, it's creating hierarchy with your design elements.

So all of these things are really important when you're trying to communicate, and that's what the lab reports are doing. After you've done all this work to get all this incredible data, you want to make sure that it's received and understood by your target audience. A professionally designed report can help convey that data by attracting interest. The data's all in there, but you want to pull out certain things that are key elements. You are also thinking about the fact that some people are not going to read the whole report, every single word. So a graphic designer can pull out the big-picture key messages you want to reveal. Design can also give different feelings by using different imagery to create a feel of the things that have come out from the various workshops.

Another thing that we do is keep a branded look and feel for the reports out of the lab. They are not all the same, but they appear in the same family so that there's an instantly recognizable branded piece. When someone sees the lab reports, they instantly say, "Oh, I know this is from this Lab." I know them. They provide good information. You already know what to expect. So you can see things similarly. You don't have to relearn where everything is or what they're going to be about. It's about brand recognition that is from the lab that does this great work.

By keeping the reports branded the same and well-designed in a similar vein, there is consistency. If you're changing designs up all the time, then the customer never knows what they're going to get. It's a visual cue for your audience to say, okay, I know where I am, I know where to look, I know what to expect. And there may be gems in here that are new and different. The topics, themes and visuals will change and will be different for each project but the structure is enough the same that it feels familiar to your audience.

Suppose someone were to do a futurecasting project. Why should they go all the way through from taking their writeup in a Google Doc or a Word doc into making it more of a communication for whomever their final audience is going to be?

Well, because you've done all this work, you've gathered all this data. Is it just to sit there and go nowhere? No, you want to communicate and share that. So anyone who is going to read the report—if there are two documents, there's a word document, very plain, no embellishments whatsoever, versus a designed report with a cover, title, all the imagery, it's going to be more enjoyable for someone to read the one that's designed. They're going to pick up the designed version more times than not over the plain Word doc. And yes, you'll have some very nerdy scientist types who are fine reading dense, heavy, data-rich scientific manuscripts. But part of your audience is going to be those scientists but the decision makers in enterprise offices or government institutions; those are not always your data nerds. They could be someone who holds the purse strings as money-focused. They will be more apt to read a designed report and get the same information but in a different way. And so I think it's important to appeal to a wide variety of audiences because depending on who needs to read this information if it's for the future, it could be important for both types of audiences. So to have that option of a designed report is key.

Stories from the Lab

During one particularly ambitious hybrid workshop, blending in-person attendees and virtual participants, the facilitators discovered they were essentially running two events at once. The in-person group thrived on the chaos of sticky notes, whiteboards, and in person conversation while the virtual attendees leaned heavily on their digital workbooks and chat boxes. At one point, a virtual participant, who could see into the room asked, "Are they building a spaceship over there? It sounds like Mission Control."

Facilitators quickly realized that bridging these worlds required constant coordination and creativity. They set up real-time polls to sync the groups and even used a digital whiteboard to merge physical post-it ideas with virtual annotations. Frequent check-ins became a lifeline, with one facilitator joking, "I feel like I'm hosting a call-in radio show while juggling!"

In the end, the hybrid model didn't just work—it thrived. The tools weren't just bridges; they became amplifiers for everyone's voices, proving that with flexibility and preparation, even the gap between physical and virtual spaces can spark future-shaping collaboration.

Chapter Resources

Vocabulary

Technical Action Report
A comprehensive written document capturing the findings, opportunities, flags, and gates from a Futurecasting workshop, including context, research foundation, and raw data appendices.

Executive Summary
A concise, high-level overview of the most critical insights, tailored for busy audiences who need the "what" and "why" without the full detail.

Briefing
A short, focused presentation of findings—often in person or virtual—designed to communicate key takeaways within 10–60 min.

Presentation Deck
A structured series of slides delivering a visual and narrative overview of Futurecasting findings, opportunities, and actions.

Graphic Novella/Comic Book Output
A highly visual storytelling format that translates speculative futures into engaging illustrated narratives.

Podcast/Audio Roundtable
An auditory format for sharing findings conversationally, often incorporating interviews, stories, and expert commentary.

Science Fiction Prototype (SFP)
A fictional yet research-grounded narrative used to illustrate a possible or potential future scenario.

Audience-Centric Delivery
Designing the output format, tone, and style around the specific needs, preferences, and expertise level of the intended audience.

Strategic Relevance
Ensuring that outputs align with organizational goals, enabling direct application of insights to strategy and decision-making.

Resources

Primary Delivery Options

- Technical action reports (full findings, context, raw data).
- Academic papers (deep dives into one or two findings for scholarly audiences).
- Executive summaries (1–2 page concise overview).
- Briefings (short verbal presentations with visuals).
- Presentation decks (8–10 slides max for clarity).

Alternative and Creative Outputs

- Graphic novels or comics (visual, narrative-driven engagement).
- Podcasts or audio interviews (accessible while multitasking).
- Digital interactive reports (clickable, multimedia-enhanced documents).

Design and Communication Support

- Professional graphic design for reports to improve clarity, hierarchy, and brand consistency.
- Branding guidelines to maintain a recognizable, trusted "look and feel" for all outputs.
- Data visualization tools to enhance understanding and retention.

Recommendations

1. **Start with Audience Analysis**—Identify who will use the findings, their context, and preferred communication style.
2. **Link Findings to Action**—Show not just what's possible, but how the audience can act.
3. **Vary Depth by Format**—Use full reports for comprehensive data and summaries or briefings for decision-ready overviews.

4. **Incorporate Storytelling**—Use vignettes, SFPs, or visuals to make futures relatable and memorable.
5. **Use Multiple Formats**—Pair a long-form report with a short summary, presentation deck, or creative format for wider reach.
6. **Highlight Key Elements Visually**—Use pull quotes, sidebars, and infographics for skimmable insights.
7. **Maintain Brand Consistency**—Ensure all outputs feel like part of the same family for recognition and trust.
8. **Test Your Output**—Share drafts with a small, diverse audience before final release to ensure clarity and engagement.
9. **Leverage Creative Media**—Consider podcasts, comics, or short videos to broaden accessibility and emotional resonance.
10. **Plan for Accessibility**—Ensure formats are accessible across devices and for individuals with visual or auditory impairments.

Useful Links:

- Multi-Sided Dice link Dice Shop: https://www.thediceshoponline.com/specialist-dice/multi-sided-dice-sets.html
- Applied Future Lab: https://www.appliedfutureslab.com/
- Threatcasting Community: Threatcasting.ai
- Threatcasting Lab at ASU: https://threatcasting.asu.edu/

Wrap of Section One

We began this book with an ontological exploration of Futurecasting, defining it as a method, a framework, and a process. Now, at the end of this section, we revisit this discussion with the added context of practical application.

For many first-time readers, Futurecasting often starts as a method. The structured phases, novel activities, and unconventional ways of thinking about the future can feel like an uphill climb—what one student aptly described as a "tough hill to climb." Yet, once you've reached the summit, you'll see Futurecasting for what it truly is: a flexible framework and an iterative process, offering opportunities well beyond the bounds of its formal method.

Futurecasting as a Framework

Now that you've navigated the method, consider how Futurecasting functions as a framework to empower analysts. Moving beyond the rigidity of structured phases and activities, the framework allows room for adaptation and innovation.

The Futurecasting Foundation, particularly its application areas, often serves as the starting point for this flexibility. Analysts may find themselves adjusting elements of the workshop on the fly to better address the unique demands of their application areas. This adaptability, while second nature to experienced practitioners, can initially unnerve workshop participants.

As participants grow comfortable within the structure of the method, introducing flexibility may feel like breaking the rules. They may worry they're "doing it wrong" when an analyst encourages them to take advantage of this wiggle room. Here, the analyst's role as a facilitator is critical—coaxing, encouraging, and reassuring participants to lean into the creative space the framework offers.

The payoff is significant. A flexible framework enriches the raw data generated during workshops, strengthening the analyst's post-analysis starting point and ultimately producing more robust findings and actionable outputs.

Futurecasting as a Process

The latest advancements in Futurecasting lie in its evolution as a process, particularly within the post-analysis phase. Practitioners have begun integrating tools from other methodologies, enhancing its capabilities as a dynamic and adaptive process.

For example, hypothesis-driven approaches have been introduced to analyze data against existing academic literature or real-world applications. This method provides an external yardstick, helping to control for variations between analysts that might otherwise lead to divergent findings. As a human-centric process, Futurecasting benefits from these tools, which mitigate subjectivity and enhance rigor.

It's important to remember that data isn't the final goal of Futurecasting. The real value lies in the models and scenarios it generates, which uncover blind spots and inspire new ways of thinking about the future. Futurecasting systematically guides participants through thinking exercises, pushing them beyond the constraints of past paradigms.

With Great Power ...

Futurecasting is still relatively young as a strategic foresight method. Its strength lies in its adaptability as an applied tool. As a practitioner, you now have the knowledge and skills to envision a range of possible futures and, most importantly, actively shape them for the better.

Futurecasting is more than a method. It's a mindset—a framework for exploration, a process for discovery, and a call to action. With these tools in hand, you're not just imagining the future; you're creating it.

Reference

Brown A (2017 Apr 18) Why should practitioners publish their research in journals? Retrieved from https://researchforevidence.fhi360.org/practitioners-publish-research-journals

Part II
The Futurecasting Method Applied

Without leaps of imagination, or dreaming, we lose the excitement of possibilities. Dreaming, after all, is a form of planning.

—Gloria Steinem

Now that a better understanding of the fundamentals of the Futurecasting Method has been realized in Part I, Part II of this book will detail the Futurecasting method as an applied process. Futurecasting can be applied and used in multiple ways, as discussed in Part I. Each time it starts with the Futurecasting Foundation. It is essential for the analyst(s) to have completed the Futurecasting Foundation before moving further. Specifying the topic and research question will give more detail to the Application Areas. For this section of the book, specifying who will use the Futurecasting and how the output will be used will determine what type of Futurecasting Workshop will be required.

What will you get from the applied section of this book?

As the title implies, the second part of this book explores the applied applications of the Futurecasting Method. This book's applied section is to be used as a "how-to" reference guide.

It can be read traditionally, one section after the other (to better understand application similarities and differences), or used as a reference guide for specific needs while pulling together a Futurecasting workshop. Feel free to move around and find the sections needed. The assumption is that you have already read Part I of this book before moving on to the applied section.

Start by asking:

- Who will use the output?
- Is this for an organization, corporation, or academic institution?
- How many participant(s) will be involved?
- Will it be a large group or a smaller group of individuals?

Knowing the Application areas will help the analyst(s) through the phases, interactions, and output of the Futurecasting performed. It will also help determine who the participant(s) will need to be. There are multiple ways the Futurecasting Method can be applied.

Types of Futurecasting

Solo Futurecasting

Futurecasting can be an individual activity; this requires comprehensive knowledge of the Futurecasting Methodology, experience as a lead analyst and connections to other analysts to verify your findings.

Individual Futurecasting is a solo dive where the analyst doubles as the sole participant, navigating through the intricacies of the methodology independently. In an Individual Futurecasting session, the analyst-participant has the autonomy to tailor prompts and modify EBMs to derive richer data. This process aids in pinpointing the specific intersections that will be explored during Futurecasting. The framework enables the creation of multiple EBMs, each visualizing a distinct scenario: a person, in a specific setting, encountering a problem.

There are advantages of the solo approach, most notably—efficiency. With only one individual involved, coordination efforts are minimized. Flexibility is also an added advantage due to the solo nature of the exercise which grants greater control over timelines. The analyst can swiftly adapt, entertain new inputs, and iterate on the Futurecasting process, producing more nuanced findings.

There are also limitations and pitfalls when going it solo. The primary drawback of this method is the inherent constraint of a single perspective. In contrast, group sessions, be it small or large, foster discussions, idea exchanges, and collaborations. Such interactions often birth perspectives that an individual might overlook. Working in isolation might deprive the analyst of collaborative insights and the opportunity for idea validation. Sharing findings and soliciting feedback can be a way to mitigate this limitation.

Solo Futurecasting is most appropriate for specific problem exploration. Analyst(s) often resort to this method when they aim to dissect a particular problem. The insights derived can subsequently inform group Futurecasting workshops. After group workshops, individual sessions can be employed to delve deeper into topics that emerged as outliers during group discussions.

It's imperative for the analyst to have a clear vision of the expected output application, considering the final audience. Solo sessions can be susceptible to scope creep due to the absence of external checks and balances. Thus, analysts must remain vigilant and anchored to the foundational principles of Futurecasting. Distractions can also creep in, emphasizing the need for focus. However, one of the beauties of this approach is its iterative nature. Analysts can always revisit and recalibrate their research questions, leading to diverse or even contradictory results in subsequent sessions.

Finally it is important to have a plan for validation of findings including working with other analysts to review final work.

Demo Prototype Futurecasting

A medium-sized Futurecasting of up to 25 participants, uses only existing data, no SMEs are interviewed, it is only a half day, in-person or virtually, and creates mockup examples of EBMs per each small team. Finally, a Demo Prototype Futurecasting produces no raw data for post-analysis and no outcome application is provided.

Amidst the diverse approaches to Futurecasting the Demo Prototype version of Futurecasting is uniquely designed to offer a succinct and illustrative experience of the Futurecasting process, without delving into the intricacies of the research phase or post-analysis. A feature of Demo Prototype Futurecasting is the participant count - this method is tailored for a medium-sized audience, accommodating up to 25 participants. The data sources used in the exercise relies exclusively on pre-existing data, negating the need for fresh data collection from SMEs or other sources.

The Demo Prototype Futurecasting is a concise exercise, typically wrapped up in half a day, making it suitable for participants with time constraints. The session can be conducted both in-person or virtually, offering flexibility in its delivery. Instead of crafting comprehensive EBMs, participants create mock-up examples in their respective small teams, they still report out to the room but typically the data is not captured. In the spirit of a demo or prototype this approach does not produce raw data suitable for post-session analysis nor does it culminate in actionable outputs or applications.

The primary aim of this method is to serve as a demonstration or prototype of the Futurecasting process. It is experiential in nature, allowing participants to gain a tangible understanding of the methodology without diving deep into the research and analysis phases. This approach is ideal for audiences keen on grasping the foundational concepts of Futurecasting without committing to a full-fledged workshop.

Sprint Futurecasting

Among the diverse strategies in the Futurecasting toolkit, Sprint Futurecasting is a streamlined and efficient approach, ideal for teams acquainted with the methodology and seeking rapid insights. This format is designed for a small, cohesive group, with a maximum of 9 participants. Sprint Futurecasting is best suited for internal teams. Public attendance is typically not encouraged to maintain the session's focus and agility.

The methodology often relies on pre-existing data, minimizing the need for extensive new data collection. True to its name, the Sprint is concise, typically lasting half a day, ensuring participants derive maximum value in minimal time. Offering versatility, the session can be conducted both in-person or virtually. While results are generated, they aren't presented in a formalized or designed format. However, raw data is made available to the sponsoring organization for further exploration. Sprint Futurecasting is especially beneficial for groups that have previously engaged in a Futurecasting exercise. Familiarity with the methodology enables participants to navigate the process more efficiently and derive actionable insights quickly. This approach is perfect for teams seeking a quick turnaround without compromising on the depth of analysis.

Rapid Futurecasting

Within the variety of Futurecasting methodologies, Rapid Futurecasting is an intermediate and structured approach, designed to deliver formalized results in a relatively short time frame. Rapid Futurecasting is designed for a medium-sized audience, accommodating up to 30 participants.

This approach integrates the insights of 3 to 5 Subject Matter Experts ensuring that the Futurecasting process is informed by experts, and additional research is gathered for a Rapid Futurecasting as well.

This type of Futurecasting is designed to span a full day, maximizing the depth of exploration within a limited timeframe.

Adapting to the sponsors preferences, the session can be conducted either in-person or virtually. Each small team within the session crafts two EBMs. A facilitator works with participants to explore potential futures. While the methodology generates data, the volume is less compared to more extensive Futurecasting approaches. Sponsors are provided with raw data for further study.

Rapid Futurecasting delivers a pre-determined application outcome. The results are not only written in collaboration with a Sci-Fi writer and a professional designer, ensuring intrigue, clarity and coherence for the intended final audience.

Rapid Futurecasting is best suited for groups seeking a structured yet swift exploration into potential futures. The integration of SMEs ensures that the exercise is grounded, while the application outcomes provide clarity.

Big Tent Futurecasting

A large group workshop of up to 75 participants uses up to 9 SMEs, is held on location over two days, creates four EBMs per small team and produces a lot of raw data for post-analysis. The outcome application provided is formally written and designed with a socialization plan.

The term Big Tent has a unique connotation. Stemming from its political origins, where a "Big Tent" platform represents a space inclusive of diverse people, issues, and viewpoints, the term has been adopted to describe a type of workshop at the Lab. These workshops, known as "Big Tent" events, are grand in scale and are characterized by their comprehensive suite of activities conducted before, during, and after the actual event.

The Big Tent Futurecasting Workshop goes beyond merely assembling individuals to discuss potential futures. Instead, it mirrors the experience of gathering a diverse cross-section of a community—ranging from local leaders and educators to industrialists and artists. Under this expansive "tent," participants collaboratively engage in exercises aimed at the collective benefit. Their involvement is purely voluntary, with no financial compensation. Yet, the voluntary nature of their participation often results in remarkably expansive and forward-thinking outputs. The vast spectrum of perspectives converges to examine a wide array of potential futures, culminating in a rich dataset ripe for subsequent analysis.

Each of the types of Futurecasting gave different advantages and disadvantages. Each of the options will also give the analyst(s) varying degrees of flexibility, volume of raw data, and final output.

Benefits of the Big Tent Approach

The larger the group, the more comprehensive the Futurecasting outcomes. This vast pool of insights proves invaluable for organizations seeking to understand potential futures and align their teams toward a unified vision of the future. Beyond the primary research findings, these workshops serve as an effective team-building and prioritization exercise.

Challenges of the Approach

Larger groups necessitate intricate planning, particularly concerning scheduling, venue management, and workshop setup. With a high number of participants, every activity demands a longer duration.

Managing a multitude of personalities can be challenging, making the role of facilitators crucial.

Implementation and Structure When tackling complex, multifaceted issues that demand extensive data and varied future scenarios, the Big Tent Futurecasting Workshop emerges as the preferred approach. Ensuring diversity is paramount. Participants must represent a mix of backgrounds, age groups, genders, and areas of expertise to ensure the broadest range of potential futures and data generation.

In terms of scale, these workshops can accommodate up to 75 participants. However, with meticulous curation, this number can be stretched to 100. For such expansive groups, the presence of seasoned Futurecasting analysts and adept facilitators becomes indispensable.

An Interview

Introduction to Applied Experiential Futures: An Interview with Kim Larkin
Founder, Experience Curator, Strategic Futurist

Could you tell us what inspired the creation of AEF and what it's all about?

Kim Larkin: Before we dive in, I'd like to start by sharing that Applied Experiential Futures (AEF) is a method I've been developing in collaboration with Cyndi Coon, the founder of the Applied Futures Lab. Cyndi's vision and leadership have been instrumental in shaping this approach. Together, we've created a methodology that blends creativity, foresight, and human-centric design to engage with the future in a deeply immersive way.

AEF emerged from the realization that while existing forecasting methods like Futurecasting and Threatcasting are powerful, they often miss a key dimension: the deep human experience. With AEF, we wanted to address that gap by creating a hands-on, dynamic approach to Futurecasting. It's about making the future tangible, actionable, and relatable—emotionally, intellectually, and creatively.

How does AEF differ from methods like Futurecasting and Threatcasting?
Kim Larkin: Great question. Futurecasting and Threatcasting are structured, analytical approaches to envisioning and preparing for potential futures. They focus on strategy and prevention. AEF takes those methods further by incorporating deep human experiential design. It's like an "expansion pack" for the frameworks, adding elements of storytelling, play, and sensory immersion to help participants not just think about the future but actually feel and experience it.

Can you give an example of how this plays out in practice?
Kim Larkin: Of course. In one session, we might use a time travel simulation where participants physically explore a future scenario. This could involve sensory props, soundscapes, or interactive storytelling. For example, in one workshop, we used symbols of nature—a river, mountains, and trees—to physically map out ideas and build timelines. These tactile elements make the abstract concrete, allowing participants to step into future possibilities and experience them in a visceral way. It bridges the gap between analysis and action.

What are some of the key features that define the AEF methodology?
Kim Larkin: AEF is grounded in a few core principles. Starting with human-centric engagement, we focus on adaptability, ingenuity, and collective wisdom. Sensory-rich experiences help participants deeply connect with future scenarios. We also incorporate immersive storytelling and play, using gamified environments to spark creativity and solutions-oriented thinking. It's a fun and effective way for adults to engage with complex challenges and bring together diverse perspectives to enrich the process. We collaborate with experts, creatives, and participants from various fields to explore future possibilities in a dynamic and inclusive way. The process produces tangible outcomes—such as prototypes, visual stories, or actionable strategies—that participants can apply directly to their work or lives.

Where do you see AEF being most impactful?
Kim Larkin: AEF is incredibly versatile. It's especially effective in sectors like education, technology, healthcare, and social innovation—fields where creativity and adaptability are crucial. For instance, to help organizations anticipate future health crises, build resilience in systems, and explore innovative responses to emerging cybersecurity threats.

And how do participants typically respond to the methodology?
Kim Larkin: They often describe the experience as transformative. By engaging the senses and emotions, AEF helps them connect with the future in a way that feels real and actionable. They leave with more confidence, creativity, and a sense of agency in shaping what's to come.

Looking forward what's next for AEF?
Kim Larkin: I believe AEF will continue to evolve as a tool for empowering individuals and organizations to embrace change. The world is becoming increasingly complex, and

AEF provides the mindset and tools needed to navigate that complexity creatively and collaboratively.

Any final thoughts you'd like to share?
Kim Larkin: Just this: The future isn't something that happens to us—it's something we create. AEF invites us to step into the role of active participants in shaping that future. It's a journey of imagination, exploration, and collaboration, and I'm so proud to be working on this with Cyndi and the Applied Futures Lab.

Chapter Resources

Vocabulary

Application Areas

The defined domains, sectors, or contexts where Futurecasting outputs will be applied, influencing participant selection, methodology type, and final deliverables.

Solo Futurecasting

A single-analyst Futurecasting exercise, used for targeted exploration, rapid iteration, or deep dives into specific problem areas.

Demo Prototype Futurecasting

A short, medium-sized workshop (up to 25 participants) using only existing data to provide a hands-on introduction to the Futurecasting process without generating post-analysis outputs.

Sprint Futurecasting A half-day, small-group session (up to 9 participants) using pre-existing data to generate raw outputs quickly, typically for internal teams already familiar with the methodology.

Rapid Futurecasting

A one-day, medium-sized workshop (up to 30 participants) incorporating SMEs, targeted research, and formalized outputs written with a Sci-Fi writer and professionally designed for a specific application.

Big Tent Futurecasting

A large-scale workshop (up to 75–100 participants) designed for maximum diversity of perspectives, producing a high volume of raw data, multiple EBMs per team, and a fully designed outcome with a socialization plan.

EBM (Effect-Based Model)

A structured narrative model used to visualize a person in a specific place addressing a problem in future, providing a foundation for backcasting and action planning.

Scope Creep

The gradual expansion of a project's objectives or boundaries without intentional planning, often a risk in solo or loosely structured Futurecasting formats.

Socialization Plan

A strategic plan for sharing and embedding Futurecasting findings across target audiences to ensure adoption and action.

Resources

Core Inputs for Applied Futurecasting

- Completed Futurecasting foundation (topic, research question, application areas);
- Pre-existing or newly collected research data;
- Subject matter expert (SME) input where applicable;
- Facilitators and analysts with methodological expertise;
- Physical or virtual collaboration tools (whiteboards, workbooks, digital platforms).

Format-Specific Requirements

- **Solo:** Experienced analyst, validation plan with peer analysts.
- **Demo Prototype:** Pre-selected prompts, small-team mock-up templates, no post-analysis workflow.
- **Sprint:** Pre-existing data, streamlined EBM templates, raw data handoff process.
- **Rapid:** SMEs (3–5), full-day facilitation plan, Sci-Fi writer collaboration, professional design support.
- **Big Tent:** Large venue or hybrid infrastructure, diverse participant recruitment, multi-day agenda, experienced facilitation team, post-event data management process.

Recommendations

1. **Anchor Every Application to the Foundation**—Begin with a clear topic, research question, and application area to ensure relevance and focus.
2. **Match Method to Goals**—Choose the Futurecasting type that aligns with available time, resources, and desired outputs.
3. **Be Clear on Output Expectations**—Before the session begins, decide whether the end goal is inspiration, raw data, or a designed report with an action plan.
4. **Balance Efficiency and Perspective**—Solo formats are fast but narrow; larger groups produce richer insights but require more coordination.
5. **Leverage SMEs Strategically**—In Rapid and Big Tent formats, SMEs add credibility and depth; integrate them at points of maximum impact.

6. **Plan for Validation**—Even in shorter formats, create a plan for peer or SME review to strengthen findings.
7. **Mitigate Scope Creep**—Use time-boxed phases and clear milestones to keep workshops on track.
8. **Design for Engagement**—Whether in-person or virtual, prepare tools and facilitation techniques to keep participants active and contributing.
9. **Ensure Diversity in Big Tent**—Recruit participants across demographics, sectors, and perspectives for maximum breadth of ideas.
10. **Prepare for Data Volume**—Larger formats will generate extensive raw data; allocate resources for post-event analysis and synthesis.

Useful Links

- Multi-Sided Dice link Dice Shop: https://www.thediceshoponline.com/specialist-dice/multi-sided-dice-sets.html
- Applied Future Lab: https://www.appliedfutureslab.com/
- Threatcasting Community: Threatcasting.ai
- Threatcasting Lab at ASU: https://threatcasting.asu.edu/

The Project

A project within the Futurecasting methodology serves as a structured exercise designed to explore and unlock opportunities within potential futures that may emerge over the next decade. By leveraging interdisciplinary collaboration—drawing on social sciences, technical research, cultural history, economics, geo-politics and narrative building—the project demonstrates the process's effectiveness in enabling participants to identify actionable opportunities. It acts as a dynamic sandbox, offering hands-on practice to validate that the methodology. Through this approach, analysts refine their ability to backcast strategies, transforming future possibilities into actionable, impactful outcomes.

The Project allows you to practice the concepts and tools introduced in Part I. By working through the prompts, synthesizing data, and applying the Futurecasting methodology, you'll develop an Effect-Based Model (EBM) and build critical analysis skills.

Throughout Chap. 8 we will insert "The Project" when it's time to "Take Action" and practice by walking the process. Over this chapter each Take Action will support building your project.

Hosting a Futurecasting Workshop: An In-Depth Guide

8

You need imagination in order to imagine a future that doesn't exist.

—Azar Nafis

8.1 Introduction

A Futurecasting workshop is a structured exercise aimed at forecasting potential future scenarios by analyzing present conditions and emerging trends. These workshops are designed to help participants envision multiple future possibilities and plan accordingly. This section serves as a guide to the process of hosting a Futurecasting workshop, either in a physical or virtual setting, to ensure a productive and insightful experience for the analyst(s), facilitators, participants, and sponsors.

Workshop Formats:

Futurecasting workshops can be tailored to accommodate different sizes and types of groups. A typical format is the Rapid Futurecasting workshop, which is a compact version ideal for medium-sized groups of up to 25 participants. This format encourages collaboration and a focused exploration of the prompts.

Hosting Platforms:

Futurecasting workshops can be conducted either physically or virtually, depending on the preferences and logistical considerations of the participants and the host.

Physical versus Virtual

B. D. Johnson and C. Coon, *Futurecasting*, Synthesis Lectures on Threatcasting,
https://doi.org/10.1007/978-3-032-21993-0_8

Physical Workshops:

Conducting the workshop in a physical setting allows for direct interaction among participants, which can lead to a more engaged and collaborative experience. It's easier to read physical cues and have spontaneous discussions, making the process more dynamic and fluid. In person also expands the ecosystem of experienced Futurecasters.

Virtual Workshops:

Virtual workshops provide a viable alternative when physical gatherings are not feasible or preferable. They require a virtual meeting platform such as Zoom, Microsoft Teams, or Google Meets, and it's advisable to have IT support on standby to address any technical issues. Pre-recorded videos and pre-event materials should be distributed before the workshop, and all expectations regarding the participants' time commitment should be clearly communicated.

It's recommended to schedule the virtual sessions in 1–2 h blocks over a week to prevent digital fatigue and allow participants more time to engage with the material.

Regardless of the format, the host should ensure that the agenda is well-structured and the objectives are clearly defined. It's also important to facilitate a multidisciplinary participant roster to ensure a broad range of perspectives and expertise.

Advantages of Virtual Workshops:

Geographical barriers are eliminated, allowing for a wider participant pool.

No travel required, saving time and resources.

Flexible scheduling can accommodate participants' availability.

Ability to work asynchronously on certain tasks.

Challenges of Virtual Workshops:

Increased effort required from the facilitator to keep participants engaged.

Coordinating schedules over an extended period can be challenging.

Potential for distractions due to a remote setting.

Whether hosted physically or virtually, a well-organized Futurecasting workshop can provide invaluable insights into potential future scenarios. By considering the format that best suits the needs and constraints of the participants, the host can facilitate a successful and enriching Futurecasting experience.

Before You Begin

Before you begin this process, start with these questions.

- What problem is the organization trying to solve?
- What will be done with the results?
- What is the timeframe?

The opportunity could modify the Futurecasting foundation or the details for how the analyst(s) conducts the workshops, so start with those questions first. If you can answer those three questions, you are ready to begin.

Project Documentation Plan

Begin your project by creating a "Project Documentation" tracking spreadsheet. Initially, as the lead analyst(s), you might create this spreadsheet in an analog format. However, promptly convert it to a digital format to facilitate easy sharing and updating. Once you finalized the list of event participants you'll add their details to the tracking spreadsheet. The spreadsheet will become host to several tabs, one of them as an example will be called "SEAT MATRIX." This matrix will act as a dynamic tool to track the progression of participants from potential attendees to invited and eventually confirmed ones. Additionally, the seat matrix is instrumental in maintaining a record of the diversity among participants.

It's important to note that even with a small team and straightforward logistics, establishing a comprehensive project documentation plan is vital. Keep in mind that sponsors or core team members may encounter IT restrictions that could impact the choice of collaboration platforms or sharing services. Therefore, it is advisable to develop a data protection strategy early in the project. This strategy should encompass plans for backups, versioning, and other relevant data protection measures.

8.2 Phase 0

Preparation and Curation

Formulating the Futurecasting Foundation:

For any size or type of Futurecasting workshop, it's vital to establish a Futurecasting Foundation which will serve as a roadmap guiding the analyst(s) through each step and decision-making juncture.

Selection of Topic:

Identifying a focal issue, opportunity or problem is the first step. This chosen topic will shape the language for invitations, workbooks, and imagery for the workshop. It will also dictate the methodology application and the intended outcomes.

Constructing a Research Question:

Once the topic is set, crafting a precise research question will help narrow down the focus. A vague or overly broad research question may hinder the effectiveness of the Futurecasting method. Begin with exploratory questions, delving into the "how" and "why" of the topic. Reflect on the significance of the topic and why it should matter to others. Pinpoint one or two intersecting questions that invite further exploration. Assess the clarity and focus of the research question, ensuring it's specific yet open-ended, necessitating a deeper analysis rather than a simple yes/no answer.

Determining Application Areas:

The insights derived from Futurecasting should have clear applications. They could inform actions by the sponsor or other stakeholders, and anticipate events, technologies, or changes over the forthcoming decade. Establish how the data collected will be utilized—whether it will be shared raw, formulated into a technical report, designed into a primer, featured in articles, podcasts, or used to create speculative narratives. Understanding this beforehand lays another stone in the foundation.

Finalizing the Foundation:

Before proceeding, ensure the Futurecasting Foundation is well-articulated and the intended use of the collected data is clear. This foundation will act as the cornerstone for the entire Futurecasting workshop, helping maintain a clear focus and structured approach throughout the process.

Laying a well-thought-out foundation is crucial for the success of a Futurecasting workshop. By selecting a relevant topic, crafting a focused research question, and determining the application areas, you create a structured pathway that will guide the analyst(s) and participants through a fruitful Futurecasting exercise.

8.2.1 Develop Futurecasting Foundation

The Futurecasting Foundation serves as a scaffolding for any Futurecasting workshop. Its specificity and clarity directly influence the guidance provided to analyst(s) throughout each phase of the process. Whether the workshop is a Big Tent or a more intimate Sprint setting, the meticulous formulation of the Futurecasting Foundation remains consistent. This foundational element is pivotal, ensuring that every step and decision aligns cohesively with the overarching objectives, regardless of the scale of the Futurecasting exercise.

Topic

The topic serves as the first foundational pillar. It is the overarching theme or subject that sponsors, researchers, and analyst(s) wish to delve into. Such topics can span a wide spectrum, from broad societal concerns.

The selection of a topic offers dual benefits. First, it provides a structured starting point for in-depth research. Second, it enables clear communication about the project's core focus to various stakeholders—be it additional researchers, participants, or supervisory committees.

Several factors can influence the determination of a topic, including the intended application of the research findings and the specific areas these findings might impact, termed as "Application Areas" in the Futurecasting methodology. Once the topic, research question, and application areas are articulated, researchers might occasionally revisit the topic to ensure its alignment with the overarching objectives of the project.

In its most rudimentary form, the topic encapsulates the primary subject or domain the researchers aim to probe. External entities, like sponsors, academic supervisors, corporate departments, or governmental bodies, might also influence or assign the topic, especially if they are identified within the application areas.

Before progressing from the topic to formulating the research question, it's imperative for analyst(s) to undertake preliminary research on the chosen subject. This involves understanding existing studies, recognizing parallel or related domains, and gathering comprehensive background information. Such preparatory work not only enriches the research process but also aids in refining subsequent stages, such as prompt curation and participant selection.

Ready to explore your topic exploration for the project?

Take Action: Topic Exploration

Use this space to note background research, investigations, and explorations conducted for a deep dive into the topic.

- Write down your preliminary notes.
- Make a list of some broader related or ancillary topics. This will help you build your research question.
- Write your topic in your project documentation notes.

Research Question.

Design Your Research Question

Once the topic area is selected, the research question aims to narrow the focus to something that can be successfully explored through a Futurecasting exercise. If the research question is too broad, unclear, or not specific enough, the Futurecasting Method will not yield the optimal results. It is important not to try to "Boil the ocean" which is an idiomatic phrase that means to undertake an impossible task given resources available. It is derived from the literal aspect of boiling the ocean, which is an impossible task given the size of Earth's waters. To avoid boiling the ocean, projects should be clearly scoped within the realm of the possible. For a Futurecasting exercise, the small team participants will explore the range of possible and potential problems and opportunities with a greater depth so a tight research question ensures a successful outcome.

What Is a Research Question?

It is the center your research and it should be:

- Clear: it provides enough specifics that one's audience can easily understand its purpose without needing additional explanation.
- Focused: it is narrow enough that it can be answered thoroughly in the time/space allowed.
- Concise: it is expressed in the fewest possible words.
- Complex: it is not answerable with a simple "yes" or "no," but rather requires synthesis and analysis of ideas and sources prior to composition of an answer.
- Arguable: its potential answers are open to debate rather than accepted facts.
- Specific: A research question should be specific but also leave room for complexity.

A research question is considered intentional when it is deliberately formulated to address a specific aspect or problem within a broader topic. This implies that the question is crafted with a clear purpose and aim, reflecting the focused inquiry and understanding of the subject matter. An intentional question is designed to guide the research process effectively, ensuring that the investigation stays on track and addresses the intended subject matter in a meaningful and productive way. It aligns with the topic, guiding the direction and scope of the study, and is formulated with consideration of the available resources, time constraints, and the desired depth of analysis. An intentional research question is one that is purposefully and thoughtfully constructed to explore a specific aspect of a topic, reflecting the researcher's deliberate aim to uncover, understand, or solve a particular problem or question within the topic and application area.

Explore

For your Foundation Research Question—Start by asking explore questions. Ask open-ended "how" and "why" questions about the general topic. Consider the "so what" of the topic. Why does this topic matter? Why should it matter to others? Reflect on the questions that have been considered. Identify one or two questions that are engaging and that could be explored further through research.

Evaluate

Next, determine and evaluate the research question. What aspect of the more general topic do you want to explore? Is the research question clear? Is the research question focused? Additionally, a research question must be specific enough to be well covered in the resourcing (i.e., time, space) available. Is the research question appropriately complex? Questions shouldn't have a simple yes/no answer and should require research and analysis.

Sample Research Questions

- Unclear: Is AI going to eliminate jobs?
- Clear: How might the workforce shift due to the integration of AI into business?
- Unfocused: What is the effect on the environment from climate change?
- Focused: How is glacial melting affecting penguins in Antarctica?
- Too Simple: How are doctors addressing heart disease in the U.S.?
- Appropriately Complex: What are common traits of those suffering from heart disease in America, and how can these commonalities be used to aid the medical community in prevention of the disease?

Ready to write your research question for The Project?

Take Action: Write a Research Question

Start by asking "Explore" questions and open-ended "how" and "why" questions about the topic. Identify one or two intersecting questions that are engaging and could be explored further through research. Cyndi likes to call these the "Constellation" around the question. Creating these points to form a constellation makes a much richer question.

An exquisite constellational of questions looks like this.

What is the future of ______________ and its impact on ______________ and the effects on __________.

Example:

What is the future of AI's impact on college students and academia as a business model?

Next, determine and evaluate the research question. What aspect of the topic should analyst(s) and participants further explore? Is the research question clear? Is the research question focused? A research question must be specific but also leave room for complexity. Reminder—never have simple yes/no answers and should require research and analysis.

Write a sample research question based on your topic in your project documentation notes.

Stop Here

Don't move forward until you can clearly write and speak your research question. Once the topic area and research question are selected the focus can be tailored to allow the exploration of the range of possible and potential opportunities and problems with a greater depth. A focused topic and research question assures that the right data and perspectives will be developed.

Application Areas

The topic and research question can be altered in multiple ways depending upon the desired application area. The application areas are the third component of The Foundation. Once you've decided upon an application areas you may need to return to the topic to make sure it sufficiently captures the nature of the project.

The output of Futurecasting identifies not only actions that can be taken but also events, technologies, and changes that could happen over the selected time horizon that will indicate whether we are moving toward enabling opportunities or away from potential problems occurring. Now is the time to determine what the data collected during the exercise will be used for. Will there be artifacts, stories and narratives, articles, podcasts, or sci-fi prototypes? Knowing this now will help build the strongest foundation possible.

Ready to build your project documentation for The Project?

Take Action: Build your project documentation

Where did you choose to document your project? Google, Word, Excel?

In your project documentation notes you should now have written down:

- What is your topic area?
- What is your research question?

Let's now add your desired application area. Also add your notes on how you determined your desired choice for your application area.

Ready to build your application for the project?

Take Action: Build your application

In your project documentation notes write down:

- How will the results be put to use?
- How might others challenge the argument?
- What kind of sources will be needed to support the argument?

Review the topic and research question to see if they will inform the application areas.

If needed, refine each of the three areas to be as concise as possible. The output of Futurecasting identifies not only actions that can be taken by the sponsor and other parties but also events, technologies, and changes that could happen over the time horizon to indicate whether you are moving toward enabling opportunities or away from the potential problems occurring. Now is the time to determine what the data collected during the workshop will be used for. Will the raw data be turned over to the sponsor, will there be a technical report, will there be articles, podcasts, or sci-fi prototypes?

Knowing this now will help build the foundation.

Stop Here

Don't move forward until you can clearly write your topic, research question and what the data will be used to create in your project documentation notes. Review all of them together one more time. Are they in alignment?

Now let's talk for a minute about who will use your outcome, findings, or product. Every organization will have different requirements and needs to apply the results of your research and findings. Understanding who will use your results should always modify how the outcome is presented.

Ready to add your application outcome recipients for the project?

Take Action: Application Outcome Recipients

In your project documentation notes doc:

- Create a target list of organizations that could be your target "Who" for your application outcome results.

8.2.2 Assemble the Teams

Core Team

The core team for a Futurecasting exercise spends a lot of time together so get the right fit or folks from the beginning and make sure everyone on the core team and commit to the required planning session. The core team should include the Futurecasting analyst(s) as well as a representative from the sponsor, and audience or users in the Futurecasting Foundation's application areas. If the Futurecasting Method is being used for a corporate, industrial, military, or governmental use, the core team could include key team members who will use and/or champion the findings. The core team will also need a support staff member to manage logistics and communications requirements.

A Futurecasting core team is typically between three to five people. If more people are added, team communications management and scheduling can be too cumbersome for the analyst(s). The Futurecasting core team makes all the decisions (including logistical decisions) and also signs off on the final output together. The duration of the entire project can take from 3 months to one year. The core team will work through all the workshop details, including the topic selection, research question, application, invitations, subject matter experts, participants, logistics, and deliverables. This team is also essential for serving as stewards and stopgaps for problems and early reports.

Ready to build your core team for The Project?

Take Action: Build your core team

- Create a list of who should be included on a core team. Do this by using character descriptions, titles or job descriptions of a type (e.g., decision-makers, Futurecasting coordinator, special advisors, sponsor coordinator), not specific people's names at this point.

The core team members are the people who will work closely with you throughout the project. Their inputs can help fill in the analyst(s)' experience or social network gaps.

Make a Short List

Once you have the preferred structure of the core team, determine the individuals you want to invite to join the core team.

- Now create a list of specific names to ask. Write down those names.
- If you don't have contact information write down names of people who might.
- Create an email invitation to send to the potential core team member, including what you are asking them to do on your core team. Provide previous examples of Futurecasting work you have participated in if applicable and if anyone else committed to be on the core team share their name, after requesting permission.

Write FAQs

Write your Frequently Asked Questions list to send via email. These make everyone feel like they know what time to show up, what to where and what to bring for snacks. There is help on this later on but try a go on your own first to see how you do.

Ready to write your Frequently Asked Questions for The Project?

Take Action: Write your FAQs

- Write a list of possible Frequently Asked Questions you could imagine coming up.
- Research online common event FAQs.

Build a Schedule

Determine the core team meeting frequency, such as bi-weekly until right before the event when you may consider moving to weekly.

Ready to build a schedule for The Project?

Take Action: Build a schedule

- Set a standard time for each meeting. This will assist the core team members to protect that time on their schedules.
- Add this meeting to the calendar and send invites to your core team for the duration of the project.
- Predetermine the meeting platform such as conference call, Zoom, Skype, Teams, etc.
- Predetermine who the host is and who the co-host is in case of that person's absence.

Steering Committee

A steering committee for a Futurecasting exercise is not essential but can be helpful; it might only be one to two people. This team is in addition to the core team. The steering committee does not work on logistics, but instead supports the core team by reaching out to their personal and professional networks. The steering committee brainstorms with the core team and asks questions. The steering committee then reaches out to their vast networks to invite SMEs and participants. Steering committee members are chosen for each workshop uniquely. They work in a variety of domains and have a willingness to share their network with the core team.

Ready to build a steering committee for The Project?

Take Action: Build a steering committee

- Create a list of who should be included on a steering committee. Note, one way to start your list is to ask the core team to open their contact lists and offer to make introductions to the individuals most connected to the research topic.
- Craft the email you will use to the connected individuals. The focus of the email is to ask if they would be willing to meet on "an all teams" call regularly to discuss the event. You will also want to give them a heads up that these are the types of questions you would like them to support the core team on.
- Reach out via email to potential steering committee members.
- Write an agenda for each of the meetings of the monthly steering committee. Include things like discussing dates of the event, considerations for potential SMEs, and participants' names. Send this in advance so they understand the ask.
- Host meetings with the Steering Committee.

Types of questions to have prepared for meetings with your Steering Committee

- Is the Futurecasting Foundation sufficient?
- Who else should be on the core team and steering committee?
- What would be good prompts?
- Who would make good candidates for SMEs?
- Who would be good participants for the workshop?
- Where should the results be socialized?

Analyst(s)

The lead analyst is the person who will be in charge of the execution of the Futurecasting exercise. They will work with the core team to set the Futurecasting Foundation and select the prompts for the workshop. The lead analyst doesn't have to be the facilitator, but often they fill that role as well. Most significantly, the lead analyst will run the post analysis and be the lead author on the output.

Facilitator

Select your facilitator early on in the process. Often, the lead analyst fills this role. If the number of participants swells, the need for more than one facilitator may arise. The Futurecasting facilitator's role is to keep the exercise on schedule, explain the details of the method, how to use the workbooks, and encourage and prompt the participants. The facilitator has worked with the core team and lead analyst for months before the kick off of the event. The facilitator for the Futurecasting exercise will get to spend more time working with the participants because there are fewer people with which to interact. This means that the facilitator can adjust and change how they provide details to the participants on the fly if needed. The facilitator is in charge of keeping the room's time, keeping the participants on task, and making the experience feel like a fun and creative exercise—constantly encouraging the participants to push their imaginations. A facilitator should have a dynamic personality and a boldness that allows for maintaining control of the floor (regardless of whether the floor is in physical space or virtual space). Throughout the entire workshop, the facilitator should ask participants frequently if they need anything or have any questions; they will and continuously connecting with them makes the process easier for all involved. The facilitator should frequently remind the participants that they are the right experts, and that's why they were invited to the workshop.

The facilitator's role becomes more challenging if conducting the workshop in a virtual environment. The facilitator serves the same function to explain the method, answer questions as well as encourage and prompt the participants. In a virtual workshop the facilitator or member of the core team will need to manage the virtual meeting space, taking on tasks like sending the small teams to their breakout rooms or meeting areas. Also, the facilitator will need to jump into these breakout rooms to make sure the participants have what they need. When conducting a virtual workshop, the workbooks should be a shared document whenever possible (e.g., Google Sheets or Google Forms) This will allow the facilitator to monitor the participants work and intervene where needed. If the virtual session has more than five breakout rooms consider adding a second facilitator for the project.

Ready to select your facilitator for The Project?

Take Action: Select your facilitator

- A list of characteristics ideal for your Futurecasting Facilitator. What type of personality would you like to work with?
- Once you know the type of personality you would like to work with next you'll want to write down a list of where to find professional facilitators. Not sure where to look? Ask your core team and steering committee if they have quality, trained facilitators in their networks.

Seat Matrix

The following are Step-by-Step Instructions provided to support the building of your seat matrix. You will revisit your seat matrix several times during the Futurecasting such as when forming small teams. If you didn't previously create tabs or unique sections in your Project Documentation you will now want to do that.

Ready to build your seat matrix for The Project?

Take Action: Build your seat matrix

- Choose the Software: Select a software program like Excel, Google Sheets, or any other preferred application to build your seat matrix in.
- Build the following tabs in your spreadsheet:
  - Create a Seat Matrix Tab. Title this tab "SEAT MATRIX." Seat Matrix Tab: Use this tab to form small groups while ensuring attendee diversity. It should include columns for the potential participants' information.

- Set Up Columns, running across the top of the sheet in SEAT MATRIX. In the first row, create columns for the following information (feel free to add or remove columns as needed):
  - Participant names.
  - Affiliation.
  - Domain.
  - Expertise.
  - Gender.
  - Generation identification (e.g., Gen Z, Millennial, Gen X, Boomer).

- Background
- Known activities or hobbies.
- Links to online bios/other media.
- Status columns (invited, registered, attending, previous attendance).
- Contact info.
- Recommended by who.

The items included on this list will be discovered from online research, and fill in participant details as they become available. Create a new row for each participant. Enter the known information for each participant as if it comes available.

Subject Matter Experts (SMEs) Tab Creation

Ready to build your subject matter expert matrix for The Project?

Take Action: Build your subject matter expert matrix

- In your chosen software, create a second tab. This tab will be your "Potential SMEs" Tab. Set Up Columns in SMEs Tab: In the first row, create columns for the following details (modify as needed, maintaining diversity):
  - Name.
  - Affiliation.
  - Generation identification (e.g., Gen Z, Millennial, Gen X, Boomer).
  - Gender.
  - Domain.
  - Background notes.
  - Bio/links.
  - Status columns (accepted, recorded, transcribed).

Populate your potential SMEs sheet with a wish list, fill it in at the beginning not with specific names but with ideal avatars of people you'd like to find. Enter your desired SMEs and any available information about them. Seek assistance from the core team and steering committee for contact leads.

Communication Plan

Developing a clear and effective communications plan is crucial for the success of a Futurecasting Workshop. This plan plays a key role in managing both the planning and execution phases efficiently. It is used to organize the core team meetings and to maintain project information, such as notes and templates. Regularly reviewing the communications plan in every core team meeting is not necessary, except when changes need to be made to the pre-established communication methods.

Ready to build your communications plan for The Project?

Take Action: In your project documentation, when building your communications plan, which should be another tab in your project documentation spreadsheet, follow these steps:

- Use the Communication Plan tab to develop a schedule detailing the frequency of information sharing with different groups, including the core team, steering committee, Subject Matter Experts (SMEs), participants, and the post-analysis team.
- Include a column to record the preferred communication methods for each member of the core team. If steering committee members or SMEs have specified preferred communication modes, make sure to note and use those.
- Add a column to assign individuals responsible for communicating with each of these groups. This helps in setting clear expectations and enhances your effectiveness in communication.
- As you continue planning your event, add more columns and fill in with relevant information, further detailing your communications plan.
- In your communications plan include all links to templates such as for email invitations for ease of accessing examples.

Stories From The Lab

Our Project Documentation Spreadsheets are our "GO TO" hub for the analysts and the core team to visit for the latest updates. We keep them open and visit almost daily during the run of a project. We also share at every meeting of the core team and with the steering committee. Building a communications plan was essential for us to manage Futurecasting Workshops. Our plans always outline how often information will be disseminated to the group (both with respect to the core team, the steering committee) and who will be responsible for these communications. This really helps the Lab set the general expectations and allows the analyst to be a more effective communicator.

8.2.3 Select and Gather Prompts

For the Futurecasting Workshop, the core team, steering committee, and analysts collaborate to gather relevant research prompts. These prompts are crucial in addressing the topic and answering the research question. They employ the STEEP structure, which stands for Social, Technological, Economic, Environmental, and (GEO) Political, to scan developments in the external macro environment. This approach aligns with the Futurecasting categories: social science research, technical research, cultural history, economic projections, trends, and data informed by Subject Matter Experts (SMEs). These categories are comprehensive, encompassing a broad range of interrelated issues, and are presented in the form of prompts.

The selection of prompts is an integral part of the Futurecasting Workshop. It allows the core team or analyst(s) to guide the prompts towards the specific application area. The number of prompts should be sufficient to thoroughly cover the problem space being explored. It's important not to limit the scope at the start. Begin with a flexible approach, perhaps starting with three to four prompts and adjusting as needed.

Research Categories

To build a list of comprehensive research categories and trends, consider these areas for creating prompts:

- Social science research.
- Technological research.
- Economic projections.
- Environmental research.
- Cultural history.
- Geo-political trends.
- SMEs data with an opinion.

Ready to build your select and gather prompts plan for The Project?

Take Action: Select and Gather Prompts

- Assign the analyst or one core team member the responsibility of collecting research prompts and reporting back during bi-weekly core team meetings.
- Conduct a brainstorming session with the core team to determine various categories or "buckets" for the research.

Subject Matter Experts

In the Futurecasting Workshop, the process of gathering and synthesizing research inputs is crucial for effective outcomes. Subject Matter Experts (SMEs) play a pivotal role in providing tailored, up-to-date prompts for the workshop. They are selected for their specific expertise and perspective on the topic. Interviewing SMEs generates content that is not only used as prompts but also in the post-analysis phase. While SMEs are beneficial for small group workshops, they are not always necessary due to time constraints in organizing their contributions. However, their insights are invaluable and should be incorporated if possible.

Creating prompts involves considering various categories, including social science, technical research, cultural history, economic projections, and trends. An ideal number of prompts should cover the entire problem space without unnecessary restrictions. Depending on the workshop size, the approach to selecting prompts might vary.

Identify attributes of an ideal SME, such as their knowledge, field of expertise, publications, lectures, and unique perspectives. Craft a communications to invite SMEs, focusing on their thoughts about the research question and what concerns them in relation to the topic. Advising SMEs that their input will inform a workshop focusing on the future and asking them to suggest key considerations for participants. Schedule to Record the SMEs' contributions with good audio quality, followed by transcription and review.

Not all prompts need to be from SME interviews. Analyst(s) can draw from various materials like online videos, book reviews, podcasts, and articles. These alternative sources provide a rich pool of information that can serve as effective prompts.

Key Actions:

- Document these notes in your Project Documentation.
- Gathering plausible prompts from diverse sources, such as online lectures, book overviews, and podcasts.
- Creating summaries or overviews from multiple sources to present to participants in concise formats like written briefs or presentations.

The goal is to provide participants with enough information to fuel their discussions and help answer the research question effectively in their small teams.

Ready to gather your SMEs for The Project?

Take Action: Gather your SMEs

Write down a list of attributes an ideal SME might have, document these notes in your Project Documentation.

- What is the ideal expert's knowledge of the topic?
- What field does the expert work in?
- Has the expert written a book or articles about the topic? If so, what are they?
- Is the expert a lecturer on the topic?
- Bonus Point: Does the expert offer a unique or opposing view from another expert you are highlighting?

Ready to write a request to your SMEs for The Project?

Take Action: Write to Your SME

When reaching out to a Subject Matter Expert (SME) for your Futurecasting project, use the following guidelines to draft your email communication:

- Initial Contact: Begin by sending an email to gauge the SME's interest in participating in the Futurecasting effort. In your message, invite the SME to share their insights on the research question and ask them to reflect on what aspects of the topic are of greatest concern or interest to them, particularly what keeps them up at night in relation to this question.
- Context of Contribution: Inform the SME that their contribution will be part of a Rapid Futurecasting Live Exercise, where a group of practitioners will explore potential futures looking ten years into the future.
- Content Guidance: Request the SME to provide 5–10 key points or considerations that the workshop participants should focus on as they model futures in the domain.
- Usage of Contribution: Explain that their input will be shared with workshop participants in a private, non-public viewing setting during the exercise.
- Final Report Inclusion: Mention that the Futurecasting Lab will produce a final report featuring their contribution, including a transcript of their video. Assure them that if they prefer not to have their transcript included in the report, it will be omitted.
- Attribution: Confirm that the SME will be credited as per their preference in any materials where their contribution is used.
- Recording Requirements: Request a 5–10 min video recording from the SME, focused on the research question and relevant topics of the workshop. Emphasize the importance of high-quality audio; suggest using an external microphone for better sound quality, if available.
- Review Process: Let the SME know that a transcript of their recording will be sent to them for review and approval.

This approach ensures respectful and clear communication with the SME, outlining the expectations and the importance of their contribution to the Futurecasting project.

Sometimes, it might not be possible to interview SMEs. In such cases, remember that not all prompts have to come from video interviews with Subject Matter Experts. An analyst can utilize a range of categories for prompts, including social science, technical aspects, cultural history, economics, trends, and data with expert opinions that might typically be provided by SMEs. By drawing from these diverse categories, the analyst can compile a variety of materials and sources to create effective prompts for the participants.

The Futurecasting analyst(s) have the ability to consolidate information from various sources, crafting their own concise summaries and overviews. These synthesized materials can then be shared with participants in the form of brief written documents or PowerPoint presentations, highlighting the key points of the research. This ensures that participants have sufficient information to engage effectively in their small teams, fostering productive discussions and beginning to address the project. Based on the topic, research question and application area, the analyst has several options to explore additional resources to consider.

- Online Video Lectures: Find concise excerpts from online video content such as on YouTube, TED Talks, keynote speeches, panels, or discussions.
- Book Overviews and Articles: Look for succinct book summaries, reviews, or articles in journals like New Scientist or Journal Nature. These sources provide well-written synopsis that capture the essence of more extensive academic or scientific papers. Additionally, consider popular articles from Forbes, WIRED, the Economist, or Scientific American for in-depth insights.
- Podcasts or Audio Interviews: Select podcasts or audio interviews that are not overly lengthy, ensuring they can be easily listened to by the participants.

All of these resources should be brief enough for easy consumption during the workshop by participants for synthesis.

8.2.4 Select the Participants

Participant Selection for Futurecasting Workshops

The lead analyst and the core team work together in selecting participants for the Futurecasting Workshop. Assistance may also come from a Steering Committee. The mix of participants will include both newcomers to the method and those who have previously participated in Futurecasting events. These participants will engage in the workshop activities and contribute to the peer review of the final findings after the post-analysis by the analyst(s). It is crucial to send the final report to all participants, allowing them to share and keep a record of what they have co-created. Given the limited number of participants, careful selection is key to ensure a diverse group.

Include individuals outside the organization, system, or industry. These might include ethical hackers, science fiction writers, filmmakers, and poets, who can offer fresh insights and a unique take on humanity. The analyst will also want to consider including experienced Futurecasters. For a smooth workshop, it's beneficial to invite participants who have previously experienced a Futurecasting exercise. These repeat "Futurecasters" can provide guidance and support to new participants, as they are already familiar with the method and process. Ideally, each small team should have at least one experienced participant to facilitate and encourage the group during the workshop.

Finding Participants

The sponsor, core team and steering committees provide names of potential participants. The team then searches each name and gathers unique data to determine their expertise backgrounds, hobbies, creative practices, and other interests (these are generally from the bio links). This is important to ensure there is a diversity of imagination in the room. Imagination is an essential tool, as is creativity, when building effects based models ten years into the future.

Creating futures in a room with strangers demands a unique intersection of people coming together, requiring more than just different jobs. The bio links allow the room's curator, usually the facilitator, to find unique humans who are as different as possible. An example of this would be learning via social media that a participant also rock climbs with a club on weekends. This rock climber then gets paired with a participant who is in a weekend rock band. They work in different fields and come from different backgrounds, creating a space for them to explore all sides of themselves and use all facets of their thinking. Hobbies and whole human activities offer glimpses of creative endeavors, and that leads to the use of imagination; these people are mandatory in the workshop. If someone is a poetry fan, they are always invited.

Share with potential participants how they will contribute. Let them know that as a participant, they will be a part of the same Small Team during the run of the exercise. The Small Teams will spend time completing the workbooks to imagine possible and potential futures to build robust prototypes known as effects-based models. Lab analyst(s) will utilize the Small Teams data and models to build future findings in the final report.

Ready to write a sample participant invitation for The Project?

Take Action: Write a Sample Participant Invitation Email:

Sample Email

Hello xxxxx-

I want to invite you to an upcoming event hosted by ____________________________ ______ for the ______________________program through ___________________. We are looking at the future of ____.

We are looking at the question: YOUR RESEARCH QUESTION HERE.

The core team from ___________________ in partnership with _________________ feel you would bring a unique and valuable perspective to this exercise. We hope that you will consider participating in the event.

On the date___________________ we will conduct a Futurecasting Workshop at location ___________________. The event will bring together participants from diverse backgrounds who will participate in a structured exercise. Participants will serve on small teams to look 10 years into the future and imagine possible and potential problems and exciting opportunities the future of ____ might hold.

The process is truly innovative and has been increasingly used to anticipate complex future events. The method depends on the rich personal insights of thoughtful participants. That is why we hope you will be able to join us and contribute your unique perspectives. If not, you might have a suggestion for someone else who brings similar value to the exercise.

As a participant, you will receive all the training you need during the workshop's opening. There is no advanced homework. You need only come with your opinion, imagination and readiness to explore potential futures ______________________________. After the workshop, Lab analysts will analyze and synthesize the raw data small teams produce and will create narratives that offer direct actions to take to enable such an event. This is all packaged in a report that all participants will review before it is published.

I am reaching out to you personally to ask if you are interested in attending. I've attached the digital Save-the-Date invitation. If you have any questions, please reach out.

DATE:

TIME:

Coffee Hour and Registration 8:30–9 am

Event 9 am to 4 pm (with post-event happy hour nearby immediately following)

LOCATION:

Thank you for your consideration in attending

Warm Regards,

Your Name

Outliers and Outsiders

When building the participants list, it is also important to include those that are not inside the organization, system, or industry that is being explored. The inclusion of participants who are outliers and outsiders is not just beneficial but essential. Futurecasting, by its very nature, is about exploring the uncharted territories—to do this effectively, it's critical to step outside conventional thinking and approach problems and possibilities from angles that are not typically considered within the framework of a specific organization, system, or industry.

The reasons for including such diverse perspectives are many. First, individuals who are not ingrained in the standard practices and thought processes of a given field can offer novel insights and challenge established norms. Ethical hackers, for instance, can provide a unique viewpoint on cybersecurity and data privacy, unearthing potential futures that might not be apparent to industry insiders.

Science fiction writers and filmmakers are a group whose inclusion can be particularly advantageous. Creatives excel in imagining and constructing worlds vastly different from our current reality. Their contributions can help in envisioning futures that are not only technologically advanced but also rich in socio-cultural contexts. They can stretch the boundaries of what is considered possible or probable, thereby enriching the dialogue with imaginative yet plausible scenarios. Similarly, poets and artists bring a deep understanding of human emotions and societal dynamics. Their perspectives can ensure that the human element is not lost amidst technological and scientific discussions. They can help in framing futures that are not only technologically feasible but also empathetic and human-centric.

Outliers and outsiders act as catalysts for innovative futures thinking. They can help prevent the "echo chamber" effect where ideas are circulated within a limited group, leading to a myopic vision of the future. By integrating diverse voices, a Futurecasting Workshop can benefit from a full spectrum of ideas, from the highly technical too deeply human-centric, ensuring a more holistic and far-reaching exploration of future possibilities.

Repeat "Futurecasters"

Futurecasting Workshops offer a role of repeat participants, or "Futurecasters." These Repeat Futurecasters are of paramount importance. These workshops, especially those conducted with large groups, require in-depth exploration. Inviting individuals who have previously participated in workshops adds immense value to the process.

The value of repeat Futurecasters lies primarily in their familiarity with the Method and the overall understanding of the process of Futurecasting. They bring a level of understanding and expertise that can significantly enhance the clarity and effectiveness of the Evidence-Based Modeling (EBM) building during the raw data collection phase. Their experience allows them to serve as informal coaches on each small group, guiding and encouraging new participants to delve deeper, while maintaining a brisk pace. This is crucial in ensuring that the workshop participants feel confident about the models they are producing.

Beyond the practical aspects of their contribution, repeat participants also add a layer of continuity to the Futurecasting process. The lead analyst and core team, possibly with the aid of a steering committee, select a mix of new and experienced participants for the workshop. This blend ensures a balance between fresh perspectives and seasoned insights. When curating participants for workshops—the smaller number necessitates a more careful consideration of each participant's potential contribution.

The Repeat Futurecasters are an important part of the success of Futurecasting Workshops. Their experience, understanding of the process, and ability to foster deeper engagement among new participants make them invaluable assets.

Track Your Project Documentation

Ready to track your project documentation for The Project?

It is important for the team to track and document the participants. After the number of attendees is finalized, the next step is to place names on a tracking spreadsheet as a part of your Project Documentation. As mentioned during the Seat Matrix set up, the names are organized by domain, expertise, gender, generation, background and so on. Columns are also included for links to online bios and other media. These links are a very important tool when building small groups too.

Take Action: Track Your Project Documentation

- Jump into your project documentation now and update your participant seat matrix with confirmed names.

Small Teams

The Role of Small Teams

Small Teams are the heart of a Futurecasting Workshop. These groups, typically made up of three to four participants, are where the majority of data generation occurs. The dynamics and composition of these teams can significantly impact the quality of insights, the creativity of future scenarios, and the robustness of the raw data collected. The process of curating and managing Small Teams is a crucial step in Phase 0 and throughout the workshop.

Why Small Teams Matter

Small Teams provide participants with a focused space to collaborate and share ideas while fostering creativity and connection. By keeping groups small, everyone has the opportunity to contribute meaningfully, ensuring diverse perspectives are included in the process. Many participants note that these teams are a surprising highlight of the workshop, enabling them to engage with people they may not have otherwise encountered.

Participants often leave workshops with expanded networks, new ideas, and even ongoing collaborations. For example, researchers from different universities have launched joint projects after meeting in a workshop, and private industry representatives have forged partnerships with government agencies. These Small Teams are a powerful catalyst for cross-sector and interdisciplinary cooperation.

Structuring the Teams

Each Small Team is assigned a distinct name—often thematic and relevant to the workshop's topic. Names can be playful and rallying, like famous aircraft, ship names, or regional landmarks. This touch adds a sense of camaraderie and identity to each team, making the workshop experience more engaging.

Curating Small Teams

The composition of Small Teams should be deliberate. While some overlap in expertise is inevitable, striking a balance between different perspectives is vital:

- Diverse Expertise: Include participants from academia, industry, government, and nonprofits. This ensures varied viewpoints and richer conversations.
- Gender and Age Diversity: Where possible, aim for diversity in gender and age to reflect a broader spectrum of experiences.
- Functional Considerations: For virtual workshops, group participants in similar time zones to ease scheduling challenges for synchronous discussions.

By understanding the participants' backgrounds ahead of time, facilitators can anticipate potential gaps or overlaps and adjust facilitation strategies to guide discussions effectively. For instance if a team skews heavily toward academia, the facilitator might challenge them to think from an industry perspective or if a group has a noticeable lack of gender diversity, the facilitator can encourage participants to consider perspectives they may not naturally adopt.

Facilitating Dynamic Small Teams

The facilitator plays a critical role in managing Small Teams. By monitoring discussions, encouraging collaboration, and ensuring no single voice dominates, facilitators can help each team achieve balanced and productive conversations. The goal is to create an environment where every participant feels valued and motivated to contribute.

Timing and Logistics

For in-person workshops, Small Teams typically work together for one to two days. Virtual workshops, which may span one to two weeks, require additional planning. Scheduling considerations and time zone alignment are essential for synchronous sessions, while asynchronous components should allow teams to engage meaningfully across different schedules.

The Long-Term Impact of Small Teams

The connections formed in Small Teams often extend beyond the workshop. Participants return to their organizations with not only new insights but also a network of collaborators from other fields. These interdisciplinary relationships can lead to innovative solutions and long-term partnerships, especially when tackling complex challenges that demand "whole of industry" or even "whole of nation" approaches.

Key Takeaway

Small Teams are more than just a logistical necessity—they are the engine of Futurecasting Workshops. Thoughtfully curated and skillfully facilitated, they transform the workshop experience into a collaborative, imaginative, and impactful exploration of the future.

Ready to build your Small Teams for The Project?

Take Action: Build your Small Teams

Prepare to build your Small Teams. You can accomplish this by using a tracking spreadsheet, a whiteboard, or a large paper sheet. If needed, gather dry erase markers, sharpies, phone, and computer charging cables. Build your Small Teams just a few days before the workshop. Don't do it any sooner as people will drop out and people could be added late. This activity is usually finalized the night before or even the morning of the workshop. Build your Small Team on your spreadsheet, whiteboard, or large paper sheet using these steps:

- Create a grid of boxes equal in number to the numbers of Small Teams you will have at the workshop. As an example, 12 people would be 4 small groups of 3 participants each. In this example create four boxes with enough space to write in the three names.
- In these boxes, begin writing participants' names by their identifiers of a domain, expertise, gender, generation, and background to ensure every team has a unique and broad makeup.
- Assign a color to each identifier. For example, the military/government is blue, industry is green, academia (faculty and students) are red, and so on. Create a key to see the names appear as their unique identifier visually. This step will help you maintain clarity of diversity for the Small Team.
- Prepare to move names around for hours, if not days. It is so important to create very dynamic Small Teams, so give this action a great deal of attention.
- Once small groups are finalized, create a slide with all of the boxes, each including the small group participants' first name only and their assigned team name such as Team "Blue Boat." This slide is used during the opening presentation as a way for each Small Team to see the names of their fellow team members and what team they are on (Fig. 8.1).

8.2.5 Construct the Workbooks

The Futurecasting Workbooks are the main tool used to capture the raw data that will be needed for the post analysis. The workbooks give the participants a place to write down their thoughts and perspectives on the prompts and are a place to capture the effects-based models, which are the main output of the Futurecasting Workshop. Analyst(s) should consider the Futurecasting Foundation as well as the makeup of the participants and Small Teams when drafting the Futurecasting Workbooks. The collection tool can be a simple

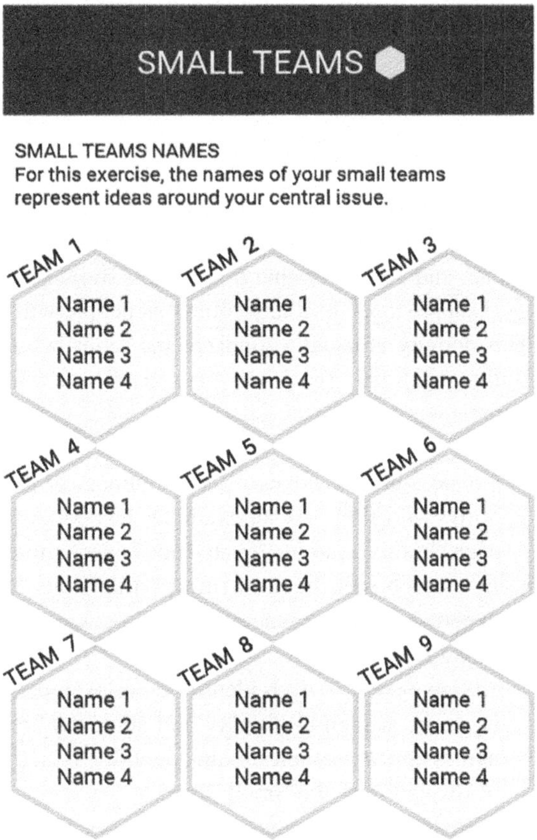

Fig. 8.1 Visual of small team names grid for a Futurecasting workshop

structured data gathering tool or a heavily designed (e.g., language, visual design) tool. Technically, the workbooks are a structured database, which is why simple spreadsheets work well for information capture. The workshop workbooks are used by the analyst(s) in the post analysis or as an input to other data processing platforms. For ease of participants' access and version control, the Futurecasting Lab generally builds workbooks in either Google Sheets or Google Forms. The formats are easy for structuring the data collection. Each Small Team has their own tab or form and by hosting the workbook on Google, each participant can easily log into the document during the Futurecasting Workshop.

Research Synthesis Workbook

The Research Synthesis Workbook (RSW) is a foundational tool used in Phase 1 of the Futurecasting Method. While Research Synthesis can be described as a method in itself, it is best understood as the starting point of the Futurecasting Workshop. This phase begins at the very opening of the workshop, setting the stage for collaborative exploration and the generation of ideas.

Each Small Team will receive specific assignments related to particular prompts, Subject Matter Experts (SMEs), and additional inputs such as videos or articles. Importantly, no single individual is tasked with synthesizing all the inputs during the workshop. The RSW ensures that the process is structured, distributed, and manageable. Participants will access the RSW using a link provided by the facilitator. Typically, this is done on laptops or tablets for ease of input and review.

Prompt and SME Assignments

The facilitator reveals each team's SME assignments during the opening presentation. Alternatively, these assignments can be shared in advance to save time. Each participant will work with specific inputs tailored to their team's objectives.

Each participant starts by individually engaging with their assigned prompts. They listen to the video or audio and reflect on the inputs provided. Participants record their observations, insights, and initial ideas in the RSW, focusing on the one or two prompts they are individually responsible for synthesizing.

Small Team Formation

Participants then transition into their Small Teams. For in-person sessions, this involves physically moving into team groups, while virtual sessions use designated breakout rooms.

Selecting Prompts for Exploration

Within the teams, prompts are selected for deeper exploration. A random selection process is used—teams can use tools like 20–60-sided dice during in-person sessions or random number generators for virtual workshops. This randomization encourages fresh perspectives and eliminates bias in the selection process.

Workbook Integration

Teams integrate their chosen prompts into their designated tab within the Futurecasting Workbook. This ensures each team has a structured and centralized space for collaboration.

Team Engagement

Teams engage in collaborative discussion to process, analyze, and synthesize the information tied to their prompts. This phase involves dynamic exploration of ideas, connecting inputs to the broader Futurecasting Foundation, and preparing for the next phase. The RSW provides a structured format that still allows creativity and adaptability. By assigning specific prompts to individuals and teams, the workload is distributed, making synthesis efficient and comprehensive. The process ensures teams are engaging in meaningful dialogue, with diverse perspectives contributing to the final output.

The Research Synthesis Phase not only helps participants become immersed in the Futurecasting Foundation but also primes the workshop for the deeper exploration and ideation that will follow. The RSW is the launchpad for building imaginative, actionable visions of the future.

For a Futurecasting Workshop, Research Synthesis Workbooks (RSWs) are essential tools for capturing participants' insights. They are designed to facilitate the construction of effect-based models (EBMs.). These workbooks, ranging from simple spreadsheets to more complex designs, are consistent in structure for both large and small group settings. The workbooks are constructed based on open-ended questions to draw out varied perspectives. Open-ended questions are included in the Research Synthesis Workbooks as a crucial way for a Futurecasting Workshop to encourage diverse and deep thinking. These questions allow participants to express their unique perspectives, fostering a rich array of insights. They enable participants to think beyond standard or predictable responses, leading to more innovative and creative ideas. This approach is essential for futurecasting, which thrives on a multitude of viewpoints to envision complex future scenarios effectively.

The Research Synthesis Workbooks are distributed just before the workshop begins to foster spontaneity and fresh thinking. Google Sheets, Google Forms, and Excel spreadsheets are generally recommended for accessibility and collaborative features. The lead analyst will prepare one Research Synthesis Workbook for each round of the workshop for participant use.

Note: A third workbook should also be prepared for the post-analysis session.

Construct a Research Synthesis Workbook (RSW), using your chosen spreadsheet software.

- Incorporate open-ended questions to structure your RSW.
- The objective is to guide participants in delving into and conversing about the given prompts, thereby processing the information effectively.
- Common questions in an RSW might be:
- Identify a data point that you consider significant or intriguing. Why?
- How might this data point influence futures or research inquiries?
- Do the consequences of this data point appear beneficial or detrimental?
- What actions should be taken in response to this data point?

Ready to build your RSW for The Project?

Take Action: Build your RSW

- Construct a Research Synthesis Workbook (RSW) in your preferred spreadsheet platform such as Google sheets, Google Forms, or Excel.
- Include the topic area and research question at the top so that it is always handy for everyone.

- Include a section for the Small Team to write or select in their assigned name if not already pre assigned.
- Incorporate open-ended questions to structure your RSW.
- The objective is to guide participants in delving into and conversing about the given prompts, thereby processing the information effectively.
- Common questions in an RSW might be:
  - Identify a data point that you consider significant or intriguing. Why?
  - How might this data point influence futures or research inquiries?
  - Do the consequences of this data point appear beneficial or detrimental?
  - What actions should be taken in response to this data point?

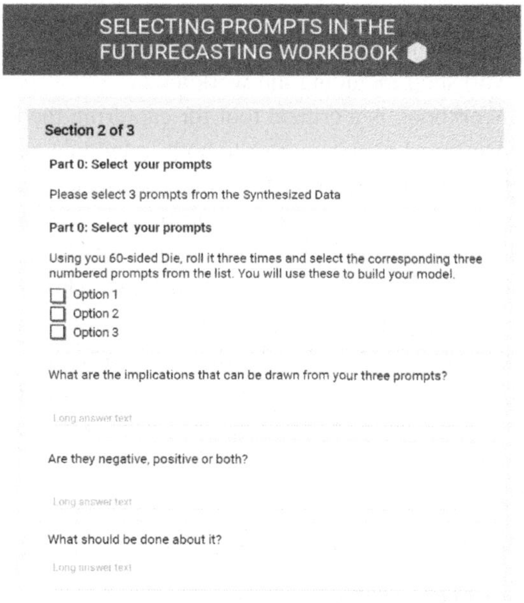
SELECTING PROMPTS IN THE FUTURECASTING WORKBOOK

Section 2 of 3

Part 0: Select your prompts

Please select 3 prompts from the Synthesized Data

Part 0: Select your prompts

Using you 60-sided Die, roll it three times and select the corresponding three numbered prompts from the list. You will use these to build your model.

- [ ] Option 1
- [ ] Option 2
- [ ] Option 3

What are the implications that can be drawn from your three prompts?

Long answer text

Are they negative, positive or both?

Long answer text

What should be done about it?

Long answer text

Figure: Phase 0 (Selecting Prompts) of a workbook

Continuing with the steps for conducting a Futurecasting Workshop, it's important to understand that Research Synthesis Workbooks (RSWs) serve as vital tools for gathering participants' insights. They aid in the development of effect-based models (EBMs) and are adaptable to various formats, from straightforward spreadsheets to more elaborate designs. Once you have designed your main RSW, create as many tabs or additional forms, duplicating the workbook, as you have small teams. Rename each to the small team names. Each should look identical at the start of the exercise. Teams will adjust their workbooks organically as they see fit—allow this!

Build Futurecasting Workbook (TCW)

For the Futurecasting Workshop, the next essential workbook to construct is the Futurecasting Workbook (TCW). The TCW is structured with a sequence of questions and tasks that guide participants through the process of designing and modeling their effects-based models (EBM). This approach leverages Experience Design techniques, enabling participants to delve into scenarios involving a person in a specific location experiencing an event. These scenarios are shaped by the prompts provided and the information gathered from the Research Synthesis Workbook (RSW) to build the EBMs.

Experience design, often referred to as XD, is a discipline focused on designing products, processes, services, events, and environments with a focus on the quality of the user experience and culturally relevant solutions. These techniques collectively contribute to creating experiences that are not only functional but also enjoyable, meaningful, and accessible to the target audience. Experience design should be built into all components of the Futurecasting Workshop including the workbooks.

The Futurecasting Workbook is a critical tool for capturing the insights of participants during the workshop. Designed to accommodate both individual and group settings, these workbooks can vary from basic spreadsheets to more intricate formats. A key feature of the workbooks is the inclusion of open-ended questions, which are pivotal in fostering diverse and deep thinking. These questions encourage participants to offer their unique perspectives, contributing to a broad spectrum of insights. This method prompts participants to venture beyond conventional answers, fostering innovative and creative ideas. Such a diverse and exploratory approach is fundamental in futurecasting, as it relies on a wide range of viewpoints to effectively conceptualize complex future scenarios.

To encourage spontaneity and fresh thinking, workbooks are distributed just before the workshop starts. Google Sheets or Forms are recommended due to their ease of access and collaborative capabilities.

Ready to build The Futurecasting Workbook The Project?

Take Action: Build The Futurecasting Workbook

- Construct the TCW workbook in your preferred spreadsheet platform such as Google sheets, forms, or excel.
- Include the topic area and research question at the top so that it is always handy for everyone.
- Include a section for the Small Team to write or select in their assigned name if not already pre assigned.

Place a note right at the top of the workbook with information for you participants such as:

- **Instructions:** *You have 2 h to complete this workbook. There are five parts. You can move backward and forward through questions and pages at any time after you've selected your team name and three prompts. Responses are autosaved as they are entered.*

PART 0: Select your prompts
In the next section "PART 0: Select your prompts" asks participants to Please select 3 prompts from the synthesized data. Include the note: *Using your 60-sided Die, roll it three times and select the corresponding three numbered prompts from the list. You will use these to build your model.*

Ready to select the prompts for The Project?

Take Action: Select the prompts

The following are the drop down long form questions for participants to answer in the workbook:

- What is the data point you keyed in on or that you found important or interesting? (Date Point).
- What are the implications of the data point on the research questions? (Implication).
- Are these implications positive or negative? (Positive/Negative).
- What should be done about it? (Should be done?) (Fig. 8.2).

Part One: Build Your Person, in a Place Experiencing an Event
Include in this workbook section the following for participants: Create your person and the specific place they are experiencing a very big event. Think about what type of future you want and what type of future you want to avoid. Describe your person using a lot of unique details to help build a realistic depiction. Make your person different from anyone on your team and lean into the details—get to know your person and really share their life.

Part Two: Expand the Experience
Construct your experience questions. Consider what questions might help the participants to tell a richer and expanded story. What would fill out the vision and give the analyst(s) more information for the post analysis process? Use the Futurecasting Foundation. The research question and application areas can give specific guidance and language for more explanation.

Use the Six Dimensions of Experience Design (Experience Questions) to explore different aspects of the experience questions (Breadth, Duration, Interaction, Intensity, Design, and Value).

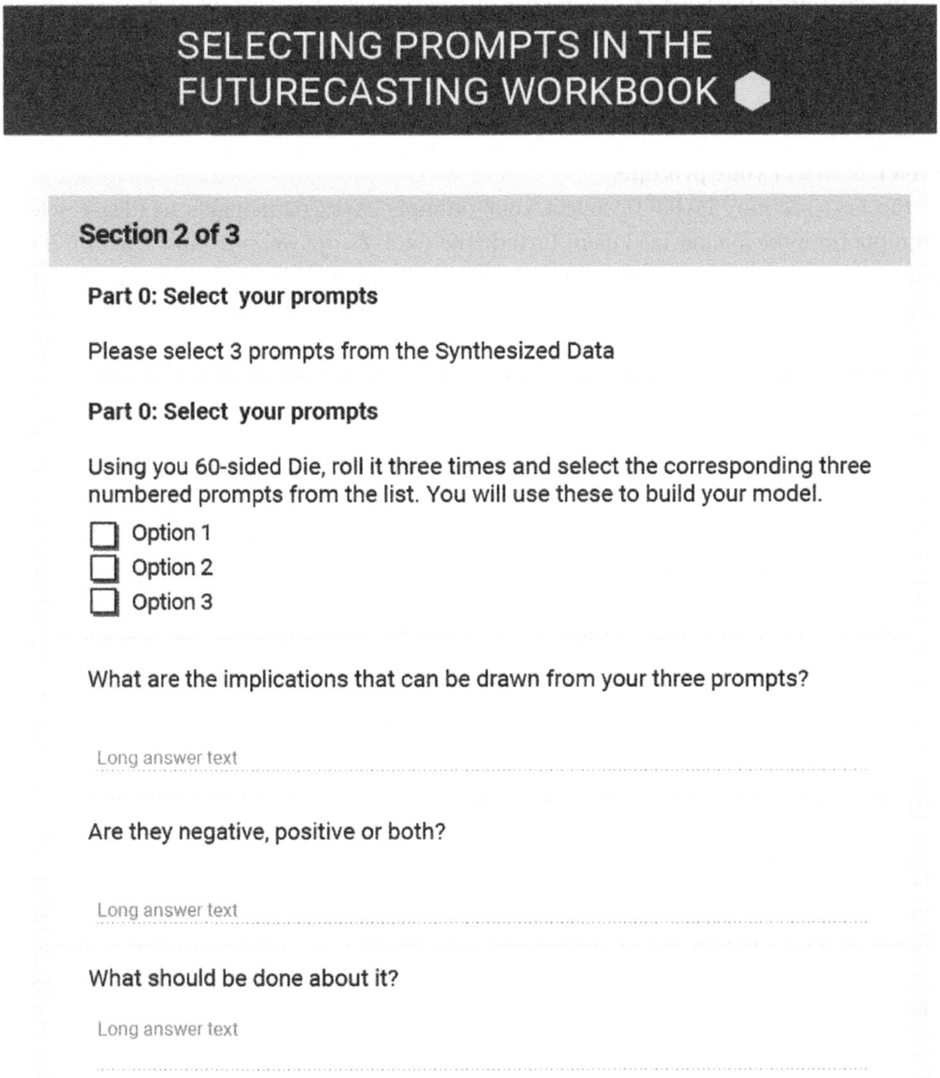

SELECTING PROMPTS IN THE FUTURECASTING WORKBOOK

Section 2 of 3

Part 0: Select your prompts

Please select 3 prompts from the Synthesized Data

Part 0: Select your prompts

Using you 60-sided Die, roll it three times and select the corresponding three numbered prompts from the list. You will use these to build your model.

- ☐ Option 1
- ☐ Option 2
- ☐ Option 3

What are the implications that can be drawn from your three prompts?

Long answer text

Are they negative, positive or both?

Long answer text

What should be done about it?

Long answer text

Fig. 8.2 Phase 0 (selecting prompts) of a workbook

Part Three: Enabling Factors

Construct your enabling questions to explore what factors helped to bring possible futures to light using the effects-based approach. The Futurecasting Foundation's application areas will give the analyst(s) areas to consider for more questions for the workbook. These questions should highlight various enabling areas or activities. Each of these enabling areas can help the analyst(s) expand and specify areas where they might want the participants to comment.

Examples of enabling questions might include:

- What economics or business need to be in place?
- Is there new research that needs to be conducted?
- Is there a specific failure of policy, law, or culture that enabled the future?
- Are there geopolitical conditions that contributed?

Part Four: Backcasting

In the section you will construct the backcasting section of your workbook for The Project, including a section to request small teams take time once they arrive at part four to reflect and discuss experience and enabling questions.

Ready to set up backcasting in the workbook for The Project?

Take Action: Set up backcasting

- In the workbook write: Teams please quickly discuss the combination of the experience and Enabling Questions as a team. Taking time to discuss what can be done to enable the future.
- Build the section for small teams to list how they might take action.
- How they will construct gates.
- How they will construct flags.
- How they will construct milestones and timelines.

Once you have designed your master workbook, create as many tabs or additional forms as possible, duplicating the workbook as many times as you have small teams. Rename each to the small team names. Each should look identical at the start of the exercise.

8.2.6 Logistics

The logistics plan for the Futurecasting Workshop is designed as a systematic series of tasks, meticulously structured to assist analysts in managing every phase of the exercise. Incorporating a dedicated tab or section in the project documentation spreadsheets is crucial for logistics to run smoothly. This tab or section will serve as a central hub, enabling both analysts and the core team to stay continuously updated and in sync regarding all activities pertinent to the Futurecasting Workshop. This organizational approach ensures streamlined communication and effective coordination throughout the workshop, laying the groundwork for a successful and productive futurecasting experience for all.

Decide Modality

Futurecasting Workshop (Physical/Virtual/Hybrid)

A critical initial step in planning your Futurecasting Workshop is selecting the appropriate modality of engagement—whether it will be an in-person, entirely virtual, or hybrid format. This decision significantly influences the logistics and structure of the event.

Physical, In-Person Workshop: Opting for a physical event means conducting the workshop face-to-face.

Virtual Synchronous Workshop: In this format, all participants engage in the workshop remotely but simultaneously. This could be structured as 3–4 h sessions spread across 3–4 days, allowing for intensive, real-time collaboration and interaction.

Hybrid Virtual Workshop: A hybrid virtual workshop blends both synchronous and asynchronous elements. Participants are all remote, but the workshop schedule includes both live sessions and periods for independent or small group work. For example, organize "All Participant" live sessions for one hour every other day, with intervals in between for smaller teams to meet at their convenience to progress with the workshop tasks.

Each modality has its unique considerations and advantages, and the choice should align with the goals, resources, and preferences of both the sponsor and the participants of the Futurecasting Workshop.

Schedule Meetings

The kickoff of the Futurecasting Workshop is marked by the core team's decision to conduct an exercise aimed at gathering data pertinent to the Futurecasting Foundation research question. The process of planning these meetings typically unfolds through conference calls or video-calling platforms, facilitating remote collaboration. However, if the core team members are in close proximity, these planning sessions can also be effectively conducted in person.

It is advisable for the analysts and the core team to establish a regular meeting schedule to ensure consistent progress and effective coordination. A recommended approach is to arrange bi-weekly calls during the initial months of planning. This frequency allows sufficient time for task completion while maintaining momentum. As the event date approaches, shifting to weekly calls in the month leading up to the event is crucial. This increase in meeting frequency ensures that all details are finely tuned and that the team is aligned and prepared for a successful Futurecasting Workshop.

Ready to assign Core Team tasks for The Project?

Take Action: Assign Core Team Tasks

- For the calls—select a core team member to be the call coordinator. This person will email either the link to the conference call number or the location of the meet-up spot to bring everyone together.

- Select a core team member to write and share the agenda call or gathering to move the project forward. Send out the agenda in advance whenever possible.
- Select a team member to be the meeting host or facilitator to keep the topics moving.
- Select a team member to be the scribe or note taker for the calls or gatherings to capture the discussion as well as capture all due outs*.
- Select a team member to follow up with any due outs after the meeting within 24–48 h.

*"Due outs" are all the things that need to get done before the next call. These tasks should always have a name attached to them so that individuals can report on progress at the next meeting.

Pick a Date

Selecting the date for the Futurecasting Workshop is a crucial step that should be addressed during the first kick-off call with the core team. Establishing the workshop date, in conjunction with deciding on the modality, lays the foundation for all subsequent planning and organization. While choosing a date might seem straightforward, it often requires careful consideration due to the busy schedules of the participants.

Typically, it's advisable to set the workshop date at least 3–5 months in advance. This time frame allows participants enough lead time to adjust their schedules and commit to the workshop. For smaller core teams and participant groups, leveraging online tools like Doodle polls or scheduling software such as Calendly can be extremely helpful. These tools streamline the process of finding a suitable date by coordinating availability across all participants, ensuring a date that works for everyone involved in the Futurecasting Workshop.

Location and Venue Selection

Once the core team has set the date for the Futurecasting Workshop, the next critical step is selecting an appropriate location and venue. The choice of venue largely depends on the chosen modality of the workshop.

For an in-person workshop, it's important for the core team to determine the geographical distribution of participants quickly. This information is essential in choosing a location that is most convenient and accessible for the majority. Additionally, the selection of an in-person venue requires consideration of space requirements. You'll need to know the expected participant count for venue size and cost. Options could include a member's workplace or even a public space, depending on suitability and comfort.

In contrast, for a virtual workshop, the "venue" refers to the digital collaboration platform to be used. Popular options include platforms like Google Meets, Microsoft Teams, Zoom, or similar software that support both large group sessions and breakout rooms for smaller team discussions. When selecting a virtual platform, it's crucial for the core team to consider any potential access restrictions that participants might face, whether they join from their workplace or home. The chosen software should be universally accessible to all participants to ensure smooth and effective participation in the workshop.

In either scenario, whether in-person or virtual, the timely selection of a suitable venue is key to moving forward with the workshop planning and ensuring a productive and successful Futurecasting experience.

Ready to build your Check List for The Project?

Take Action: Build a Checklist

This list is designed to guide the core team in evaluating potential locations and venues for the Futurecasting Workshops. Create sections for:

- Venue tracking section or tab in the project documentation spreadsheet. List venue name, contact number, website, contact person, and event package link.
  - Location considerations include What is the travel time and distance for most participants?
  - What time of year is it? This matters because those in cold climates would almost always prefer to take 2–3 days in the warmer west as opposed to vice versa. The time of year can also determine hotel costs. Beware of locations with a negative connotation for some of your participants. For example, government employees have a hard time justifying attendance at events in Las Vegas based on the nature of the city. Pay specific attention to any location restrictions that the Futurecasting sponsor has to consider by their organization. Travel considerations ensure all attendees can travel to the selected location. Assessing travel time and distance for participants is a generous courtesy.

- Cost Considerations: Determine costs for venue rental, hotels, and meals in the host city.

- Venue Capacity and Layout: Confirm adequate space for the expected number of participants.
  - Check for enough desks/chairs and arrange tables for small group settings.
  - Verify availability of a side room for calls and a quiet main room.
  - Ensure an overflow space for small group sessions, with comfortable seating.
  - Ensure the spaces are accessible to all participants.
  - Are there convenient shops and facilities for quick coffee or lunch breaks?
  - Request food allergies in advance for catering purposes.
  - Confirm nearby food options for pick up or delivery of catering.

- Audio-Visual Equipment.
  - Confirm availability of AV equipment with projection and audio capabilities. Ensure AV—screen, projector, and sound system are available in the main room.

- Wi-Fi and Power Outlets: Verify Wi-Fi availability and additional charges. Ensure enough power outlets for all participants' devices.

- Facility Rental Details.
  - Understand costs and booking conditions.
  - Check time restrictions and ensure morning access at least one hour before the event. Typically, reserve from 7:00 a.m. to 5:00 p.m.

- Venue Features.
  - Provide ample desks/chairs and alternative seating arrangements.
  - Check for IT support and a small conference room for private calls.
  - Evaluate parking or transportation options.

- Security and Protocols.
  - Review security and emergency protocols at the venue.

- Additional Fees.
  - Check for any additional fees (e.g., for Wi-Fi, AV equipment, parking).

Save-the-Date

As soon as the date, location, and modality of the Futurecasting Workshop are confirmed, it is essential to promptly notify potential participants. This early notice is crucial to allow invitees to express their interest and ensure their availability by adjusting their schedules accordingly.

There are several different options for Save-the-Date notifications, including the following as examples:

Email Announcement: An email with all the necessary details, providing an easy way for participants to ask follow-up questions.

Digital Cards: Customized e-cards designed with workshop branding, offering a personal touch.

Calendar Invites: Direct calendar invitations through platforms like Google Calendar or Outlook, setting automatic reminders for participants.

Reservation Event Page: Create an event page on platforms such as Eventbrite to engage your audience.

Physical Mailers: For a special touch, especially for high-profile workshops, sending a physical save-the-date card can be impactful.

Essential Elements to Include on a Save-the-Date:

- List the event name with the full and precise name of the workshop.
- List the research question.
- Date and time: The scheduled date(s) and time(s), including time zone information.
- Location: Venue details for physical events or platform information for virtual events.
- Modality: Clearly state whether the workshop is in-person, virtual, or hybrid.
- Call to action: Instructions for expressing interest or registering for the event.
- Contact information: Provide contact details for further inquiries or assistance.
- Visual elements: Incorporate branding elements such as logos, thematic graphics, or a color scheme associated with the Futurecasting Workshop.
- Anticipated outcomes: Highlight the key benefits or outcomes participants can expect from attending.
- RSVP deadline: A clear deadline for recipients to respond by.
- The save-the-date should strike a balance between providing enough information to pique interest and keeping details concise to ensure clarity and prompt action.

To engage potential participants effectively, develop a digital SAVE-THE-DATE postcard (digital) that resonates with the theme and purpose of the Futurecasting Workshop. The creation of this digital invitation should be a collaborative effort, with the Lead Analyst proposing the initial design concept and the core team providing their thoughts, ideas, and endorsement. Ensure that the design is crafted for easy electronic distribution, such as via email or digital platforms. For the save-the-date digital postcard, the core team should appoint a member skilled in digital graphics to execute the details.

Ready to build your Save the Date for The Project?

Take Action: Build Your Save-the-Date

- Event Title: Formulate an event title that is both intriguing and informative, encapsulating the essence of the Futurecasting Workshop.
- Visual Imagery: Choose visual elements that complement and highlight the Futurecasting Workshop's thematic focus.
- Workshop Details: Detail the date(s) and format of the workshop, specifying whether it will be in-person, virtual, or a hybrid of both.
- Research Question: Feature the central research question to anchor the Futurecasting Workshop's purpose.
- Branding Elements: Integrate logos of sponsors or hosting entities, adhering to established branding standards.
- Distribution Preparation: Ensure the postcard is in a low-resolution format ready for digital distribution to participants upon their expression of interest.

By following these guidelines, the save-the-date will not only announce the event but also establish the professional and thematic tone of the workshop, encouraging potential attendees to respond quickly with a yes!

Registration

Determining the registration approach for your Futurecasting Workshop is an important task that the lead analyst and core team must undertake. For managing participant sign-ups, you have a couple of options. An Automated Registration Platform that utilizes an online ticketing and registration service like Eventbrite, which can streamline the process, especially if you anticipate a larger audience. Or you could use a manual registration for a more personalized approach, a core team member could be assigned to take this on or the lead analyst, to personally handle the invitation and registration process. For manual registrations, this can be efficiently managed by updating your seat/participant matrix to reflect confirmed attendees. This method allows for direct communication and simplifies the process by eliminating the need for a third-party service. The registration plan should be in place before the save-the-dates are sent.

Participant Planning Information (P.P.I)

Once attendees have confirmed their participation, it is important to provide them with detailed logistics information immediately. For a Futurecasting Workshop, it's common to have a local participant understanding of location and venue. However, for those who will be joining from other locations, early receipt of logistical details is essential. This information facilitates their travel arrangements and accommodation bookings, ensuring they can plan their participation with ease and convenience.

Ready to build your PPI Checklist for The Project?

To facilitate a seamless and efficient experience for all attendees of the Futurecasting Workshop, developed a comprehensive participant planning information (P.P.I) checklist. This resource is tailored to equip participants with all the essential details needed to navigate the logistics of the workshop, catering to both physical and virtual attendees. The checklist aims to preemptively address any questions related to travel, accommodation, venue specifics, and technical requirements, thereby ensuring that participants can focus on the immersive experience of the workshop with minimal distractions.

Take Action: Build a P.P.I Checklist

Build a Physical Workshop Checklist out of the following important areas of consideration:

- Venue address: Provide the complete address for the workshop location.
- Public transit details: Offer information on public transportation routes and stops near the venue.

- Parking facilities: List nearby parking garages, including the pricing if known.
- Accommodation options: Suggest local hotels or lodging options, with contact information.
- Airport information: Identify the nearest airports for out-of-state attendees to facilitate their travel plans.
- Event schedule: Share a detailed agenda of the Futurecasting Workshop to assist participants in coordinating their travel and attendance.

Build a Virtual Workshop Checklist:

- Workshop schedule: Clearly outline the schedule, including start and end times for each session.
- Platform access: Specify the digital platform that will be used, providing login details and technical requirements.
- Technical run-throughs: Propose a pre-event tech check session for attendees unfamiliar with the virtual platform to prevent technical difficulties during the workshop.

Frequently Asked Question (FAQS)

As part of the registration confirmation for the Futurecasting Workshop, providing participants with a Frequently Asked Questions (FAQs) list is a crucial step in setting the stage for a successful event. This list clarifies what is expected of the attendees, covering a range of common questions. By equipping participants with this knowledge in advance, you'll ensure that they can arrive at the workshop at ease and fully prepared to engage in the forward-thinking activities planned.

Ready to build your FAQ Sheet for The Project?

Take Action: Build an FAQ Sheet

Specifics will be provided and may vary based on the workshop's physical location.

This FAQs sheet should be shared with participants as an immediate response to their registration confirmation to ensure they have all the necessary information for a smooth and productive workshop experience. No one observes—ever. Everyone participates. To begin the build of your Frequently Asked Questions (FAQs) Sheet for Futurecasting Workshop Participants Workshop you'll want to start with the modality. Let your participants know right away if the workshop is physical, virtual, or hybrid? If virtual, details of the virtual meeting space, including the platform and file-sharing methods, should be provided in the FAQs.

- Preparation:
  - Let your participants know if there is technical support offered such as if tech testing sessions are available. If they are what days and times for tech testing will be offered for those unfamiliar with the platform.
- Workshop Schedule:
  - Let your participants know the schedule structured to utilize participants' time effectively? Note that in a virtual setting, interaction is key, while content consumption can be done individually, the building parts should be done as a team.
- Dress Code for Physical Workshops:
  - Let your participants know what they should wear which includes comfortable, casual attire. The motto is "Flip-flops encouraged" to ensure comfort. Suits and uniforms are discouraged and we ask everyone to hang all titles at the door; those aren't needed during the workshop either.
- Team Assignments:
  - Let your participants know when they will find out their team assignments and that they cannot request specific teams. Team assignments will be revealed at the workshop. Participants cannot know in advance or choose their teams as those are curated by the lead analyst. Teams will remain constant throughout the workshop. Active participation is required—observers are not permitted.
- Breaks and Availability:
  - Let your participants know when breaks are scheduled. Breaks are planned, but participants should limit long calls to remain available to their teams.
- Equipment Needs:
  - Let your participants know what they should bring to the workshop, including a personal laptop or tablet with Wi-Fi capability is necessary. Workbook access will be provided the day before the workshop.
- Accommodations and Transportation:
  - Let your participants know what the details regarding lodging and travel are.

Equipment

For a successful Futurecasting Workshop, various types of equipment is needed to ensure a productive and engaging experience for participants. The exact requirements can vary depending on whether the workshop is in-person, virtual, or hybrid. Here are some common types of equipment that might be helpful:

- Computers/Laptops: Essential for access to digital tools, research materials, and workshop documents. Have extras on hand.
- High-Speed Internet Connection: Crucial for virtual and hybrid workshops to ensure seamless communication and access to online resources.
- Projection Equipment: Projectors and screens for displaying presentations, visual aids, and collaborative work.

- Audio Equipment: Microphones and speakers to ensure clear communication, especially in larger venues.
- Video Conferencing Tools: For virtual or hybrid workshops, reliable video conferencing software and hardware are needed.
- Whiteboards, Sticky Notes, or Flip Charts: Useful for brainstorming sessions and visualizing ideas.
- Workbooks and Writing Materials: Digital workbooks, pens, markers, and notepads for note-taking and sketching out ideas.
- Collaboration Software: Tools like shared digital workspaces for collaborative activities.
- Breakout Room Facilities: Smaller rooms or designated areas for group discussions and activities, important in physical settings.
- Charging Stations: Multiple charging ports or power strips for participants' electronic devices.
- Furniture: Comfortable seating and tables arranged to facilitate interaction and collaboration.
- Snacks and Beverages: To keep participants energized and hydrated throughout the workshop.
- Technical Support Staff: For troubleshooting any technical issues that might arise with equipment.
- Accessibility Equipment: Such as hearing loop systems or sign language interpreters, if required by participants.

Randomizing Equipment

During the phase where participants build their effects-based models using the workbooks, a mechanism to randomly select data points from the combined Research Synthesis Workbook is essential. This randomization process adds an element of unpredictability and creativity to the workshop.

For in-person Rapid Futurecasting Live Exercises, ensure that you have an adequate supply of randomizing tools such as dice or playing cards. Each small working team should have access to their own set of these tools. Dice with 30 to 60 sides are particularly useful for this purpose, as used in our lab.

In contrast, for virtual workshops, you will need to integrate a digital pseudo-random number generator. This digital tool should be accessible to all participants and easy to use within the virtual platform of the workshop. Ensuring that this digital solution works seamlessly is key to maintaining the flow and engagement of the participants during the exercise.

The following are examples of online random number generators to use when selecting prompts for the Research Synthesis Workbook.

- Websites like Random.org offer customizable random number generation, ideal for selecting data points or making decisions in a workshop setting.
- Spreadsheet Functions: Tools like Microsoft Excel or Google Sheets have built-in functions (e.g., RAND() or RANDBETWEEN()) that can generate random numbers.
- Dedicated Apps: Mobile or desktop applications specifically designed for random number generation. Many of these apps offer the ability to customize the range and type of numbers generated.
- Virtual Dice Roller Tools: Websites or apps that simulate dice rolls, some even offering various types and numbers of dice (e.g., 20-sided dice), which can be used in place of physical dice.
- Random List Pickers: Online tools that randomly select an item from a pre-entered list. This can be useful if you have specific data points or scenarios you want the participants to choose from.
- Collaboration Platform Tools: If the workshop is being conducted virtually on platforms like Zoom, Google, or Microsoft Teams there are built-in or add-on features and bots that can perform random selections.

Catering

For an in-person Futurecasting Workshop, catering can vary based on preference. The core team may choose to create a local map for participants to independently seek out coffee during breaks, which can offer a refreshing change of pace. However, providing on-site food and beverages can be a significant time-saver. Should your venue require advance catering orders, remember to include this in your project documentation tracker to ensure timely arrangements. Essential items like water, caffeinated beverages, and candy should always be on hand to keep energy levels high during the workshop. When it comes to catering it is important to schedule snack and bio breaks into the delivery times. This helps maintain participant comfort and focus throughout the event. Effective catering management, whether for physical or virtual workshops, plays a crucial role in maintaining participant engagement.

Ready to build a catering plan for The Project?

Take Action: Build a catering plan

- Accessibility: Ensure catering options are available nearby, if not on-site. Confirm any food allergies with participants in advance.
- Morning Arrangements: Have breakfast options such as pastries, along with coffee and tea, readily available.
- Lunch Considerations: Offer easy lunch options like buffets, boxed meals, or grab-and-go selections that allow participants to dine without significant downtime.
- Afternoon Refreshments: Provide a second round of coffee, tea, and snacks to re-energize participants.

- Confectionery Touch: Place bowls of candy or mints on tables throughout the workshop space for a sweet energy boost.
- Catering Setup: Arrange for a catering table within or very close to the workshop area, prepped with morning essentials and maintained throughout the event.
- Catering Contact: Designate a catering contact person to handle all food and beverage-related logistics.

No-Host After-Hours Gathering

Organize an After-Hours Social for post your Futurecasting Workshop. Following the conclusion of the day, it's beneficial to arrange an informal gathering. This social event is an opportunity for participants to strengthen connections with their small teams, fostering trust that can lead to more profound insights and community building.

For in-person events, consider planning a casual, no-host gathering at a nearby restaurant or casual spot after the day's activities. While attendance is optional and some may have personal commitments, this extension of the workshop into a more relaxed environment can significantly enhance the group's connection. The after-hours event should feel like the experience of a "summer camp" atmosphere, encouraging participants to engage in deeper conversations about the day's work and the broader implications for the community.

For virtual workshops, creating a sense of togetherness can be challenging due to the lack of a shared physical space. Nonetheless, you can host a virtual happy hour on the same platform used for the workshop. This session should be dedicated to social interaction, allowing participants to unwind and interact more personally. Ensure that the facilitator is present to lead the virtual happy hour with engaging activities and facilitate closing discussions.

Ready to build a Morning of Checklist for The Project?

Take Action: Build a Morning of Checklist

Build a Morning of Checklist

On the morning of the Futurecasting Workshop, having a detailed checklist is vital to ensure a smooth start. Here's a comprehensive list to prepare for the arrival of participants:

- Preparation: Distribute usually by email, links to the Research Synthesis Workbook and the Futurecasting/Backcasting workbook to all participants before their arrival.
- Check-In Station: Set up a welcoming check-in area just outside the event space, clearly visible from the entrance. Equip the check-in table with items like clipboard with the final registration list, writing materials such as pens and notepads, a device (laptop or tablet) for registration confirmation—ensure it's charged and has an appropriate stylus if necessary. Pre-arranged participant name tags, sorted by small team for easy collection.

- Participant Reception: Upon entry, provide participants with their name tag and hand out small team assignments symbolized by colored game pieces or mini plastic toys, instructing them to keep these items with them to use mid-day to find their fellow Small team members.
- Seating Guidance: Inform participants they may sit anywhere during the opening session; team grouping will occur later.
- Event Room Setup: Arrange tables and chairs in the event room to accommodate teams of up to four. Set up the catering area within or close to the workshop space, ensuring coffee, tea, and water are available.
- Technical Setup: Confirm that screens are operational with all presentations pre-loaded. Ensure Wi-Fi is active and accessible. Position extension cords and charging stations strategically around the room.
- Amenities: Place bowls of candy on tables for participants to enjoy.

Name Tags

Design Name Tags for your Futurecasting Workshop

For an engaging and organized start to the Futurecasting Workshop, create themed name tags that echo the inventive spirit of the event. For instance, you could style the name tags after superheroes, imbuing each one with a unique design. On each name tag, print the participant's first name prominently, along with the last initial if there's more than one person with the same first name attending. This will help prevent any confusion during the workshop.

When sorting the tags on the welcome table, group them according to their respective small teams, ensuring each set carries a similar design element. Additionally, assign a distinctive small toy or game piece to each team, with all members of a single team receiving items that are uniform in color, shape, or design. These items will not only serve as a fun icebreaker but also as a visual cue to reinforce team identity throughout the event. Display the name tags in an organized manner, making it easy for participants to find their small team when it is time.

Build Deck

Preparing the Futurecasting Workshop Deck for a Futurecasting Workshop

Construct the Futurecasting Workshop Deck well ahead of time using software like PowerPoint, Canva, Google Slides, or your preferred presentation software. Clearly outlining the agenda and activities is a crucial aspect of communication that serves to manage participant expectations. Frequent and clear information not only prepares participants for what lies ahead but also fosters a sense of familiarity and trust.

Maintain the presentation visible throughout the duration of the workshop. For in-person events, this means having the slides displayed on a screen within the room. In a virtual format, keep the presentation up in the main room, even as participants move into breakout sessions. This consistent visibility ensures that everyone can reference the current instructions or timeline at any moment, keeping the workshop on track and focused.

Ready to build a Futurecasting Workshop Deck for The Project?

Take Action: Build a Futurecasting Workshop Deck

For the Futurecasting Workshop, craft a deck that will guide participants through the event. Ensure this deck is not only informative but also visually engaging to keep participants focused and excited about the process from start to finish. The Futurecasting Workshop Deck should include:

- A Welcome Slide: Feature sponsor logos and a warm welcome message.
- Housekeeping Slide: Indicate the location of bathrooms, availability of beverages, and timing for breaks.
- Agenda Slides: Present a detailed agenda for the day. For virtual workshops spanning multiple days, create separate slides outlining daily activities, which will provide clarity on the sequence and timing of events.
- Inspirational Opening: Start with an opening poem chosen by the lead analyst or a team member to set a reflective tone.
- Futurecasting Introduction: Include slides that introduce the Futurecasting process with resources for reference.
- Key Definitions and Acronyms: Explain relevant terms and acronyms related to Futurecasting space being addressed.
- Small-Teams Reveal: Use a grid slide to display small teams, showing names and a team image. Remind teams to select a scribe and a reporter for the two-minute trailer-style report-outs.
- Research Synthesis Process Overview: Outline the steps for the Futurecasting Research Synthesis Process.
- Prompt and Subject Matter Expert (SME) Assignments: Reveal the prompts and SME assignments, guiding participants on how to engage with the Research Synthesis Workbook (RSW) for their designated prompts.
- Team Coordination Slide: Instructions for team formation, either moving physically for in-person or transitioning to breakout rooms for virtual sessions.
- Futurecasting and Backcasting Procedures: Detail the processes for both Futurecasting and Backcasting within the workshop.
- Report-Outs: Provide a slide with the structure for the two-minute report-outs, encouraging concise, impactful summaries akin to movie trailers.
- Guidance: Instructions for repeating the process with new prompts, advising teams to either randomly select or quickly choose without delay. We often call this Rinse and Repeat.
- Closing Slide: Provide space for participants to reflect and share their thoughts at the end of the workshop.

Primer

Create a Primer for your Futurecasting Workshop. A well-crafted primer is an invaluable resource for participants in a Futurecasting Workshop. It serves as both an introductory guide and a reference document throughout the event.

Ready to build a Primer for The Project?

Take Action: Build a Primer

- When designing this primer, consider the following elements to ensure it's comprehensive and user-friendly:
- Workshop Overview: Start with a brief description of the workshop's goals, structure, and the main activities that will take place.
- Futurecasting Fundamentals: Explain the concept of Futurecasting, including its objectives, benefits, and how it differs from other forecasting methods.
- Agenda and Timelines: Provide a detailed agenda with session times and durations to help participants manage their participation and break timing effectively.
- Glossary of Terms: Include definitions of key terms, acronyms, and jargon participants will encounter. This is particularly important for newcomers to the field.
- Research Synthesis Workbook (RSW) Guidelines: Offer simple step-by-step instructions on how to use the RSW, highlighting how it fits into the Futurecasting process.
- Backcasting Instructions: Remind what the backcasting method is, illustrating how to work backward from future scenarios to present-day actions.
- Small Teams Dynamics: Outline how teams will function, the roles within each team (e.g., scribe, reporter), and the importance of collaboration.
- Prompt Interaction: Guide participants on how to interact with the prompts, including listening strategies and synthesis techniques.
- Reporting Procedures: Clarify the expectations for the report-outs, the format, and tips for effectively sharing their team's report.
- Technology Use: For virtual participants, include a section on the technical aspects, such as logging into platforms, accessing digital workbooks, and troubleshooting common issues.

Registration

The lead analyst or core team will also need to make a decision on how to handle registration. An online service such as EventBrite works well or one core member can keep track of the list of registered attendings by having them email a confirmation directly to that person. The morning of the event the registration list should be printed and placed on a table at the entrance to track participants' attendance and provide them with the workshop supplies.

"In a Futurecasting workshop everyone works … everyone must be a participant and add their ideas to a small group. No one sits back and observes. There are no visitors. It is a room full of doers to support each other with their passion and participation. The sausage-making takes all hands on deck."

—the Authors

No-Host After-Hours Gathering

For an in-person Futurecasting Workshop, consider scheduling a no-host after-hours gathering at a restaurant within walking distance from the Futurecasting venue. With many of the participants (if not all) local, this may or may not be of interest to the group as many will have family commitments to get home to. However, if it does work, this will create more of the "summer camp" feeling you want participants to feel, and when participants come back the next day, their trust in each other has grown, and they do a deeper data mining of themselves and each other.

Additional Individuals

Depending on where the Futurecasting Workshop is held additional individuals may be required to be onsite. Some of the individuals to consider are:

- Event management team for overall coordination provided by the venue or the sponsor. This team oversees all aspects of the event, ensuring that everything runs smoothly and according to plan. They communicate with sponsors, vendors, and local authorities. They ensure the event stays within financial constraints and that all activities are compliant with relevant venue rules, laws, and regulations.
- Some venues may require Security Personnel to monitor and manage the movement and behavior of attendees to ensure safety and lead the response to any emergencies or unexpected situations. Ensure only authorized personnel and attendees enter specific event areas. Act quickly in emergencies, providing initial response until other services arrive. Work with local police if necessary. Manage a lost and found area for items misplaced during the event.
- Some venues may also require medical staff and emergency medical care personnel to provide immediate medical attention to attendees or staff in case of injury or illness. Operate and manage onsite first aid.
- Volunteers can offer information distribution. Assist in providing information to attendees, such as directions, schedules, and other event details. Assist with Set-Up and Tear-Down. Provide support to the event management team, security, and medical staff as needed. Help attendees with special needs or those who require additional assistance.

8.2.7 Care and Keeping of Humans

Walking Breaks

Build in breaks during the Futurecasting process. During a multi-day workshop, participants should be provided with places they can easily walk to that are filled with nature, interesting sites, or shopping. We do this because there is a lot of sitting and writing during a workshop, and people still need to move. We call these walking breaks and use them, so that participants don't get stagnant during the exercises; we encourage participants to take a walk.

Even if a workshop is virtual, breaks are important. Build breaks into the agenda and encourage the participants to get up and walk around, get some water or maybe even do some light stretching. The virtual meeting space can be left open if participants want to hang out and chat during the break.

At the end of the day they will feel exhausted because the intellectual demand placed on them throughout the day will appear physically. Giving them breaks and encouraging them to step into nature will limit their exhaustion.

Hydration

The workshop participants should be provided with water and caffeine all day. Consider providing fruit throughout the day too. They will be doing a lot of thinking throughout the workshop and staying hydrated will help keep participants from getting headaches. Caffeine will also help keep energy levels up.

Snacks

Always provide snacks all day. Consider healthy snacks such as fruit, nuts, and granola bars. It's a great idea to have candy on the table but not too many sugary carb heavy snacks—can cause energy to dip.

8.3 Workshop Opening

The opening of the Futurecasting Workshop (virtual or physical) moves very quickly. Therefore, it is important to take the time to create the creative environment that the process needs (regardless of whether the event is in-person or virtual).

At the Lab, we believe words matter greatly, and this is why we open every workshop with a poem. Our team takes turns selecting a poem and reading it at the opening as our welcome to the workshop. We also cite these poems in our reports because, to us, humanity should always come first and we utilize poetry as a stand-in for the human experience.

Ready to build an Opening Session Checklist for The Project?

Take Action: Build an Opening Session Checklist

- Select a core team member to read their chosen poem as the opener.
- Select a core team member to welcome everyone.
- The facilitator should review housekeeping items such as bathroom locations, fire exits, etc.
- A team member from the lab gives a brief history of the lab.

Workshop Agenda

Each day of the workshop should have an agenda built into the presentation to provide instructions for the participants. The lab's presentation deck includes the following items:

Ready to build an Agenda for The Project?

Take Action: Build an Agenda. Include the following elements:

- Introduction to the Futurecasting Process;
- Relevant definitions;
- The reveal of the assigned Small-Group Teams;
- Phase 1: Research Synthesis Process;
- Phase 2 + 3: Futurecasting + Backcasting Process.

Sample Agenda

Let's say that you want to start the day at 9:00 a.m. Then, you should "schedule" registration and coffee from 8:30 a.m. to 9:00 a.m. to attempt to get the stragglers in before the event starts.

Generally, the opening session takes about 15 min. Then about 45 min is spent reviewing the agenda and discussing the Futurecasting process (recall that for many of the participants, this will be all new to them). The reveal of the assigned Small-Group Teams allows participants to see who they will be working with for the duration of the workshop. Additionally, due to the diverse backgrounds of the participants, it is important to review the critical definitions that the Futurecasting workshop will tackle.

Sample Event Agenda

8:30–9:00 AM

Coffee and Registration

9:00 AM

Welcome and Introduction to the Futurecasting Process

10:00 AM

Team Formation and Transition to Small Groups

10:00 AM–12:00 PM

Exercise Round One

Small teams engage in focused discussions in breakout spaces.

12:00 PM

Working Lunch

Buffet or boxed lunch—return to the main room.

12:15–12:45 PM

Round One Report-Outs and Debrief

Team presentations and wrap-up of Exercise Round One. Introduction to Exercise Round Two.

1:00–3:30 PM

Exercise Round Two

Small teams reconvene in breakout spaces for deeper exploration.

3:30 PM

Final Report-Outs and Closing Discussion

Team presentations, process review, and next steps.

4:00 PM

Optional Local Gathering (No Host)

Informal networking at a designated location.

Once the opening session concludes, the floor is yielded to the workshop facilitator to start the show. The facilitator is in charge of keeping the room's time, keeping the participants on task, and making the experience feel like a fun and creative workshop. A facilitator should have a dynamic personality and a boldness that allows for maintaining control of the floor. Professional facilitation of the workshop will help move the event along and help ensure the quality of the raw data. Suppose the group is toward the high end (i.e., close to 75 participants). In that case, you should have two co-facilitators to assist with the myriad things that will pop up during the event (both futurecasting-related and logistics-related). These facilitators could come from the core team or hired professionals. Throughout the entire workshop, the facilitator should frequently ask participants if they need anything or have any questions; they will, and walking around, continuously connecting with them makes the process easier for all involved. The facilitator should frequently remind the participants that they are experts, and that's why they were invited to the workshop. The facilitator's role becomes more challenging if the workshop in a virtual environment.

The facilitator serves the same function to explain the method, answer questions as well as encourage and prompt the participants. In a virtual workshop the facilitator or member of the core team will need to manage the virtual meeting space, taking on tasks like sending the small groups to their breakout rooms or meeting areas. Also, the facilitator will need to jump into these breakout rooms to make sure the participants have what they need. When conducting a virtual workshop, the RSW and EBM workbooks should be a shared document whenever possible (e.g., Google Sheets) This will allow the facilitator to monitor the participants work and intervene where needed.

Workshop Presentation

Use the Futurecasting Workshop Deck. It's important to tell the participants what they will experience during the facilitated opening and then say it again as a way to over-communicate. This helps set the expectations, and the repeated messages build familiarity and trust with the participants. Keep the presentation up on the screen during the entire workshop—whether it is a physical display in a physical room or in the main virtual room when they are in breakouts. This way current directions for the task are at hand or a display of the current timeline is available for participants to track.

8.3.1 Presentation of Prompts

Present the Prompts

The prompts are presented to the participants to conduct the research synthesis activities. These prompts can be Subject Matter Expert (SME) videos or "buckets" that combine videos with other research mediums. This phase can be conducted during the workshop, assigned to participants as "pre-work" or pre-synthesis can happen by the analysts to save time in the room. If pre-work, the participants will watch, read or listen to the prompts on their own time, fill out the RSW, and come to the workshop prepared to discuss in their small team in order to capture the "wisdom of the room."

Whether the prompt viewing and individual research synthesis is conducted before or during the actual Futurecasting workshop—it is still critical to have time in their small working groups to discuss and expand their individual RSWs in order to capture the input from all participants.

8.3.2 Data Capture in Workbooks

At or just before the start of each Futurecasting workshop, all participants are provided with a link to the workbooks, a shared document space, or digital files to use to capture the data. The small teams will use these to write their ideas, notes, and details. Using workbooks allows more data to be captured than appears in the verbal report outs because no one can do justice to everything that was in their heads verbally. The idea then for the workbooks is to get as much detail written down as possible. The workbook gives the small group space to start looking at the problem from various angles. As the workshop progresses, each small group begins to populate the workbooks' fields.

Use the time to explain the workbooks and express to the Small Teams that the workbook entries should be thought of as writing a movie script. If they are encouraged to be creative and imaginative by writing a movie instead of being worried about if they know how to build an effects-based model, they will have fun and provide far more details. These details become more obvious when small teams recall they need movie-level inclusions; this is very useful for two reasons. First it gives more details for the analyst when it comes time to analyze. Second building in the favorite ice cream flavor of their person isn't important in the analysis but it is super important to get the small team to co-create, bond, and build empathy for the person they are building.

Scribes

When the small teams gather to build their workbooks (e.g., RSW, TWC), the facilitator encourages the small teams to pick a person who will serve as the team scribe. This will be the person who will capture the conversation, discussion, and disagreement that happens in the group. The scribe is essential to the Futurecasting Method. If the data is not captured fully in the workbooks, it cannot be used later in the workshop nor will it be of any use to the analyst(s) during the post-analysis. Because the scribe is so important, the facilitator should check in with each group and identify the scribe. They should be thanked and told that they are an important part of the method. The facilitator should then encourage the rest of the group to also write in the workbook, to capture as much information as possible. High-performing groups often will have multiple members of the small team add information to the workbooks. For small teams, the workshop might take place over a single or multiple days and so the facilitator should encourage the small groups' members to share the scribe task. A single person should not be the scribe for every step in the process if possible.

8.3.3 Report Out

As soon as the small teams come together, in addition to a Scribe, they should select a team Reporter. Throughout the course of the Futurecasting workshop, the small teams will "report out" multiple times to the other participants in the room. Small teams report out so that everyone in the room can hear what everybody else is thinking. During report-outs, participants have a strict time limit to present the most important thing(s) that came from their previous activity. With respect to this first "report out," the small teams are asked to treat their report outs like a movie trailer. No one likes a trailer that say it all. Often, you might hear the comment "I don't have to see the movie now—the trailer revealed all!" In Futurecasting report-outs we want to avoid that feeling. The report-outs can first provide the prompts that came from the research synthesis activity, the answers to the other research synthesis questions for those three points can also be included.

The reporter should provide:

- Their person's name.
- Their person's occupation.
- The relevant details to set the stage with the location.
- The reporter should then move to the event that occurred and the outcome.
- Closing with the opportunities enabled.

The facilitator keeps the time and to create a gamified environment, they should dramatically express "Go" and start a stopwatch on a phone, on a screen or with a timer such as an hourglass so everyone in the room can keep track of the time. When the timer goes off at the end of the allocated time (often 2 min), let the alarm go off in the room, so the reporter knows time is up. The facilitator can be playful in declaring "Time Is Up" so cutting folks off mid-sentence doesn't feel rude but instead game show host like.

Throughout the course of the Futurecasting workshop, the small teams will "report out" movie trailer style multiple times to the other participants in the room. Small teams report out so that everyone in the room can hear what everyone else is thinking. This also supports a game-like feel because after each round, when teams hear each other's report outs aka movie trailers they get excited by possibility and often things like "Oh I didn't realize you could do that!" can be heard in the room. This fuels small teams to create a model for the next round that has more details and pushes the possibilities further.

8.4 Discussion

Once all the small teams have presented for that round, it is important to open the floor to a short facilitated discussion to encourage participants to share what didn't get talked about—perhaps highlighting some of the side conversations that they had in their small groups that were not presented. "What didn't come up" is an important element of Futurecasting because what isn't talked about often illuminates the things that should be discussed. You should have an analyst or a co-facilitator take notes during each of the discussion sessions as the comments may prove helpful in the post-analysis phase.

- Include a space in the workbook to remind the small teams to take the time for this discussion.

8.5 Phases 2 and 3

Futurecasting/Backcasting Process

This phase of the Futurecasting Method uses the data and perspectives captured in the RSW to envision possible and potential futures. This futurecasting transitions from the high-level curated perspective from the prompts to a specific effects-based model (EBM). Participant(s) will develop this EBM by integrating elements of participatory design, experience design, and the science fiction prototyping process. This phase ends with a future that will then be backcasted, exploring how the future might be enabled. At its very basic level, the EBM is a structured database (e.g., spreadsheet) that poses a series of questions to practitioners and requires a list of tasks to fill in in order to create a qualitative model. To capture this data in the EBM, the Futurecasting Method uses the practice of participatory design. This collaborative nature of looking into the future encourages discussion and debate, ultimately generating more detail and data in the EBM. The questions and tasks contained in the EBM draw from a range of tools, theories, and practices. Typically, a ten-year time frame is used, but shorter or longer time spans can be used. During this section, in their small team, participants will participate in Futurecasting/Backcasting.

Ready to build out the Futurecasting component for The Project?

Take Action: Build out the Futurecasting component

For this exercise the small teams will use the workbook sent via a link earlier that day or loaded on a device for them in the room. At the lab we often preload the workbooks onto tablets and laptops already in place in the room for the teams to jump in.

The first section of the Futurecasting workbook tasks the participant(s) to enter the foundational data points (curated from the prompts) that they will use to inform the rest of the workbook.

- Prompt One____.
- Prompt Two____.
- Prompt Three____.

After establishing a mental visualization of the environment based on the prompts and RSW, the participant(s) imagine a specific person living in that future. This adds the human dimension to the EBM. By focusing on a single person, the participant(s) isn't overwhelmed by the enormity of the future they are considering. They are not tasked with worrying about "getting it wrong." They are simply exploring a single person, in a specific place, experiencing an event on a future time horizon.

Participant(s) are tasked with envisioning who the person is, their family, and the broader community with which they identify. As any good storyteller would say, no one is truly completely isolated, and so the participant(s) are asked to create the community with which their person best identifies. This community (whether physical or digital) provides them the support during their upcoming trial and will also be whom they tend to turn to.

- Who is your person: ____ _____.
- Who is their family, and the broader community: ____________________ ______________________________.
- Where does your person work: __________________________________ ______________.
- What are their hobbies: ____ ________.
- What does a day in their life look like? __________________________ ______________________.
- Where do they live? ____ _____.

These questions create the structure for a more realistic, human centric dimension. Depending upon the audience and the research goal, participant(s) can be prompted to expand details about their person's surroundings and daily life as well as their hopes, wants, desires and dreams.

The next step is for the participant(s) to describe the Event. The physical or digital instantiation of the problem caused—is the "event." To better model and understand the event, the participant(s) are asked a series of questions that are designed to push them to add as much detail as possible to explore and explain their futures.

The Event

- Describe the event your person experiences? ______________________ __________________________.
- How does your person experience the event? ______________________ __________________________.

The next section of the Futurecasting workbook draws directly from the practice of experience design. To gather detailed information for the EBM, participant(s) are asked a series of questions to expand their description of the person's experience with the solution and experience.

Build your Example Experience Questions

- What is different and/or the same as previous events?
- When the person first encounters the opportunity, what will they see? What will the scene feel like?
- What will they not see or understand until later?
- How will information be delivered to the person? Where and how will the person connect and communicate with others?
- What new capabilities enable the person and their broader community?
- What are the broader implications? What might a ripple effect look like?

Then, as in all good science fiction stories, the perspective changes and the "event" is seen from the opportunity perspective.

Build out your opportunity

- What is the opportunity?
- What are the motivations that caused them?
- Who might your person enable the opportunity?

The next in the futurecasting phase of the Futurecasting Method is to understand what forces might enable the opportunity to happen. The nature of the tasks and questions shift from experience-based to effects-based. Drawing from EBM, participant(s) are tasked to explore a wide range of factors that could enable the opportunity. These are the enabling questions.

Build your Enabling Questions

- Barriers and Roadblocks: What are the existing barriers that need to be overcome to bring about the opportunity?
- New Practices: What new approaches will be used to bring about the opportunity and how will this affect their broader community?
- Business Models: What new business models and practices will need to be in place to enable the opportunity?
- Research Pipeline: What technology is available today that can be used to develop or advance the opportunity? What future technology will be developed?
- Ecosystem Support: What support is needed? What industry/government/military/local partners must align with outcomes of the opportunity?
- Training and Outreach: 2 What training is needed to enable the opportunity?

The Final step of the Futurecasting Workbook EBM is to BACKCAST. The participants should get specific and "name names." Meaning, as the group explores the flags, gates and milestones, they should consider who is going to take this action, where they are and what their purpose is. In this case being vague is good. Even if the participants don't have the specific expertise or understanding for how an action could and should be taken, they should write down as much information as they can think of—even if it feels like science fiction. This can even be a call out that more work needs to be done.

Example vague language that can be effective:

- a. broader cultural conversation must be started to explore …,
- b. academia needs to research the possibilities of …,
- c. an industry consortium needs to be convened to define…..

Stories for the Lab

Every time we conduct a workshop we love to share with the small teams that they do not have to go in order while working through the EBM workbook. Meaning, the small team can skip around, filling in different details and information that strikes them. Giving them the option to race through the EBM as quickly as they can to get to the end, followed by the option to revise, edit, and add has been an effective strategy to get more detail and more robust gates, flags, and milestones. Allowing this moving around to happen organically lets each team build organically as pieces come to them. All the while, we as facilitators keep small teams on time for the number of sections they should have complete at certain stages in the workshops—this helps keep things moving.

8.5.1 Capture the Raw Data

When building the EBM with a person, in a place, experiencing an event—getting started is often the hardest part for the participants, especially if it is their first time filling out an EBM. Often the small teams want to pick the "right" person or get their future perfect before they begin. The facilitator should encourage them to just "jump right in." The facilitator can spend time with each team and encouraging them to give their person a name helps step into the shoes of the person they are building. Having the small team pick a specific city also helps to get them started and also narrows down the expanse of ideas. Participants can feel overwhelmed with the blank page so getting something written as soon as possible helps to break the ice.

The participants are encouraged to tell the best story they can tell. In this way, the secondary/sensing questions can give more detail (e.g., What will it smell like when the person experiences the opportunity?). Tailoring these questions right from the start, based on the Futurecasting Foundation, can expand the quality of the raw data in the EBM.

Experience Questions

Typically for a Futurecasting workshop, the analyst(s) will have crafted the experience questions in Phase 0 specifically for the Futurecasting workshop. The analyst can always tweak these questions in successive iterations of Phases 2 and 3 during the workshop if they initially don't gather data that helps to answer the Futurecasting Foundation.

Enabling Questions

Like the experience questions, the enabling questions for a Futurecasting workshop can be more specific. The aim for this approach is different from the experience questions. The enabling questions in the EBM are used to specify what has happened or what is needed to bring the opportunity about. Rooted in Effects Based Operations, the domain expertise of the participants and the curation of the small team takes on added importance in the Futurecasting workshop. Again, the facilitator should remind the participants that their professional backgrounds matter but that they should also focus on the Futurecasting Foundation's Application Areas and the use of their imagination to call up things they know and have heard of. The more specific they can get, pulling from real experiences and knowledge, the more useful the raw data set will be for the analyst(s).

Clarifying Questions

Often, asking clarifying follow up questions can help the participants think through even more that can be collected for analysis.

Examples of clarifying questions might include things like:

- Is there a specific law or regulation that was or was not passed?
- What group, industry, or market might need to develop a specific technology?
- Are there specific areas of investment that needs to happen from specific organizations or countries?
- Have you seen something like this before but applied to a different future?

Usually the participant(s) know more and have more detail in their minds than they are entering into the EBM. A few simple questions can help spark them to explain more. Get them also to document this in their workbooks.

Gates, Flags, Milestones

Backcasting is the final part of the EBM. This is the area where many small teams can get bogged down, tripped up, or struggle to finish. First, this is usually a casualty of their time management. The small teams will spend more time on the opening sections of the EBM and not save enough time to explore the gates, flags, and milestones. While small teams are in this section it is especially important for the facilitator(s) to keep a steady rotation of visits to every small team. This is true for both in person and virtually. The facilitator(s) should ask a lot of check in questions to ensure the participants understand the structure for building gates, flags, and milestones.

The facilitator(s) will also need to encourage the participants that they don't have the specific expertise or understanding for how an action could and should be taken, they should write down as much information as they can. Sometimes being vague is right and the analyst can connect dots during post analysis. These can be a call out that more work needs to be done and so should also be documented in the workbook.

By the second round of the workshop the small teams will have usually gotten better at time management. If the workshop is multi day the participants will have a very solid handle on the process by the second day.

8.5.2 Rinse and Repeat

Typically, the futurecast and backcast process is repeated two to four times during the course of a Futurecasting Workshop. Allowing for the small teams to fill out new and unique TCWs for each round. The more rounds that can be accomplished, the greater the volume of raw data for the post analysis. Unique means that each small team will select new data points from the RSW prompts for each round. Additionally, after each round of EBM creation, the lead analyst could modify some of the questions or add constraints to the participants and we frequently do this at the lab. Examples of the final round to constrain might be the location to a single state or city or a requirement that their person be of a certain occupation. This decision is left to the lead analyst.

There are many ways to slightly change the foundation of the EBM in order to get different data. Mostly, these changes are to push the small teams in a different direction from their previous EBM to ensure that enough unique data is captured to cover the required Futurecasting Foundation. Also, a reminder to keep all the EBMs straight there should be a new TCW created to use for each round. This will also help with the post analysis to understand the new constraints imposed in each iteration. And, since the small teams will remain intact, it will also allow the lead analyst to see their growth in ideas.

If the Futurecasting Workshop is held virtually, there is some flexibility in timing. For instance, if it is an asynchronous event, then the participants will need adequate time (outside of the event) to create the EBM together, so potentially leave 24–48 h between report outs. This gives them time to handle life events and provides a better chance that the small teams find adequate time in their schedules to meet. If it is a synchronous virtual event, the timings for the creation of the EBMs can be about the same as the in-person workshop.

8.5.3 Closing the Futurecasting Workshop

Bring all participants together for a final time in the room together or virtually in a single place. Use the Deck to provide participants with a quick reminder of what they just participated in. Now that the participants have given so much time and energy to the exercise be generous. Over-communicate what the participants can expect next—this will help to answer many questions. End the exercise with discussion, reflection, and gratitude to honor the participants' contributions also encourage participants to continue working with the analyst(s) during the peer review phases. Encourage team members to stay in touch—this grows the ecosystem of practitioners. If a physical event invite the participants to the no-host post gathering event.

If the workshop was physical, return the venue space back to its original condition. Clean up. Work with catering and IT to sign off and close out the paperwork for the venue. Before you lose attention of the participants check all workbooks were saved and make sure you have backups of all the workbooks and notes from the exercise. If there are any documents left in a shared meeting space, remove them. If the workshop was virtual, save the conversation, notes, and recordings. Thank everyone involved.

8.6 Phase 4

After the completion of the Futurecasting Workshop and the generation of the raw data, the next phase in Futurecasting is post-analysis.

8.6.1 Analyst(s)

Because the volume of raw data developed in the Futurecasting Workshop the lead analyst will generally want to get different perspectives as it can be helpful to bring in additional analyst(s) for review. Typically, each analyst will work alone on the three steps of post analysis: summary, meaning, and novelty. Once the three steps have been executed, the analysts convene and present the results. Regardless of how you structure your analysis cell, it is important to begin by creating a timeline for your actions.

Timeline

The analyst(s) will need to craft a timeline for the post-analysis that supports the required completion date of the final products.

Ready to build a timeline for the analysts for The Project?

Take Action: Build a timeline for post-analysis

- Coordinate with all analysts on the project on their availability.
- Determine whether to conduct the post analysis in-person or remote.
- Schedule a first kickoff review session to take place within a few days following the workshop.
- Provide ample time for the analysts to conduct their individual analysis of the raw data.
- Convene analysts together for collaborative sessions periodically through the process. This time allows the analysts to cross reference notes.
- Draft findings.
- Establish reviews of the findings with the core team and participants.
- Finalize findings.
- Present findings to the steering committee.
- Begin work on final output.

Post-analysis Workbook

The post-analysis workbook is a part of the project documentation, typically a spreadsheet. It is a place for the analyst(s) to capture each round of the post analysis (summary, meaning, novelty, outliers). It is beneficial to use this so that it can be shared with other analysts and also so that it can be included in the final output, assuring transparency in the methodology. Therefore, just like the other workbooks used throughout the Futurecasting Method, we use Excel or Google Sheets for the post-analysis workbook.

8.6.2 Prepare the Raw Data

Data

The output from a Futurecasting Workshop needs to have a data management and data protection plan to maintain the transparency of your analysis work. As part of your project documentation you will want to create folders that are organized to store the RSWs and EBMs. All of the analysts will need access to this raw data throughout the post analysis process. The lead analyst will also deliver these organized items to the sponsor of the Futurecasting Workshop.

Data Hygiene

Phase 4 starts with data hygiene on all the data in order to prepare it for the post analysis. Once the assigned team member, often a support person to the lead analyst, conducts the data hygiene step the data is ready for the analysts to begin their work. The chapter in the first half of this book in the methodology section provides further details on the following steps.

The data hygiene steps start immediately following completion of the workshop. Ensure that there is a complete copy of all the RSWs and EBMs saved in an off-line (but protected) location. This will ensure that data can be recovered if it is accidentally corrupted during Phase 4. As a part of your kick off meeting with all of the analysts will include updating them on what has been done on data hygiene since the workshop ended. Diss your data management and storage strategy with the analyst(s).

8.6.3 Perform the Analysis

Kickoff

The lead analyst has created a timeline and sent it to the team of analysts who will join the project. The lead analyst schedules a kick-off session with all analysts. At the kick off they review the schedule, walk through the locations of all resources and materials the analysts will need to access. The lead analyst walks the other analysts through the research question, observations from the workshop, areas of sponsor interest to highlight, the foundation and the steps (summary, meaning, and novelty) for analysis. After all materials are reviewed there should be a time for questions and discussion before each analyst goes off to work alone.

Typically, each analyst will work alone on the three steps of post analysis: summary, meaning, and novelty. Once each the three steps have been captured, the analysts convene and present the results to facilitate a group analysis on the raw data. Generally, this can be done over the span of 2–3 days if in person or 1–3 weeks if virtually and depending on availability of the analyst(s). If conducting the analysis virtually, each analyst can create and work in their preferred environment.

Tools

Every analyst has specific tools they prefer to use and that can be accommodated into the process. This makes it important to capture their analysis tools, techniques, and results as a part of your project documentation.

If there is only a single analyst, the lead and main analyst can review their own notes and in their project documentation write notes on what might help as additional perspective before beginning.

When the analyst(s) convene, discuss, and document how each person is thinking about approaching the raw data. Always set a timer for each analyst's solo post-analysis report. This will help everyone stay focused and on track with time. The analysts need to be facilitated too.

I shared the project documentation template so that each analyst can use the one assigned to them. This will help analysts avoid accidentally writing on top of each other inside your project documentation.

Have each analyst write down the Futurecasting Foundation on a sticky note or piece of paper so that they can refer to it throughout the process. Remind the analysts to periodically step away from the data to clear their heads. All analysts should keep notes in the project documentation for any insights, glossary terms, ideas, or early clusters to share with other analysts.

8.6.4 Generate, Validate, and Review the Findings

Generate

Following the three steps of post-analysis, the analyst(s) will generate their findings. Using the Futurecasting Foundation as a guide, the analyst(s) reviews Round 3 "novelty" and considers how they answer the Research Question as it is applied to the Application Areas.

It is important to remember that it is the lead analyst's role to have an opinion based upon the initial research and SME interviews when combined with the raw data sets from the workshop. Sometimes it is necessary to perform secondary research/interviews to flesh out the findings.

Validate

To validate the findings, the analyst(s) should first explore if the findings answer the research question and application areas. Next, the analyst returns to the initial research and SME interviews to validate the findings and also identify if there are any outliers or missing perspectives. Integral to the validation of the findings for a Futurecasting workshop is communication with the core team, steering committee (if it is being used), participants, and SMEs.

8.6.5 Communication

Core Team and Steering Committee

It is important to keep the doors of communication open between the analyst(s) and the core team/steering committee during the post-analysis. Confirm with the various core team members how they wish to be involved in the post-analysis. Some core team members are curious and will appreciate updates throughout Phase 4 and others prefer to see the draft results at the end of Phase 4. Also, the analyst should let the steering committee (if you are using one) know when they expect to see raw results of the post-analysis for their feedback—schedule update meetings with the Core Team and Steering Committee. Ultimately, ask these groups to conduct a peer review of the initial findings and to also sign off on the final report before it goes public.

8.6.6 Participants

It is also important to maintain communications with the workshop participants during the post-analysis phase. First, give them the opportunity to share any closing thoughts about the venue, process, and event. Next, ask them how they want to be attributed to the final products. Some participants are fine with listing their names with organizational affiliation, and others prefer to remain anonymous. Also, many participants will have multiple affiliations, so this is the opportunity to determine which one they want to use next to their name. Finally, the participants should be given an opportunity to serve as peer reviewers on the findings.

8.7 Traditional Outputs

Building the Applied Outcome

In Futurecasting, the process of converting findings into actionable outputs is crucial. This phase is not just about providing future findings; it's about effectively communicating what these findings mean to inform decision-making. Analysts need to consider their audience's needs and how the information will be practically applied. Technical reports are a common output, offering a detailed and systematic presentation of Futurecasting findings. These are essential for stakeholders requiring in-depth analysis and a clear understanding of potential future scenarios and their implications.

Some possible examples of outcomes include Science Fiction Prototypes (SFPs) which are narratives used to explore and present possible futures. These narratives, often engaging and imaginative, can stimulate innovative thinking and broaden the discussion around future possibilities. Graphic novellas and comic books translate these futures into visual narratives. This format makes the information accessible and engaging, particularly effective for audiences less inclined towards traditional analytical reports. Podcasts and interviews offer an auditory approach to discussing Futurecasting insights. This format appeals to audiences who favor conversational and auditory forms of information processing, broadening the reach of Futurecasting findings.

Each of these outputs serves a specific purpose and caters to different audience preferences, ensuring that the insights from Futurecasting are communicated effectively and are actionable. The choice of output should align with the target audience's needs, ensuring that the future insights are not only conveyed but are also positioned to inform and guide action.

Technical Reports

The staple of outputs from a Futurecasting workshop (no matter the size) is a professionally written technical report. Typically deliverables to show the results from using the Futurecasting Method. As you are working through the phases, keep this output type. This way you can think about the graphics and photos you might want to incorporate into the final technical report as well as which of the participants' EBMs you might want to turn into narrative form as part of the report's design.

Ready to build a Report for The Project?

Take Action: Build a Report

- Create an outline framework for the final report.
- If there is more than one analyst or writer, divide up initial writing responsibilities.
- Write the first draft.
- Select graphics to be used that will enhance the report sections.
- Select participants' EBMs to use to highlight a finding.
- Turn those EBMs into a short narrative.
- If more than one writer is present, have each writer review/edit the other sections in order to create the second draft.
- Share this second draft with participants and members of the core team for their feedback and comments.
- Review the provided comments and incorporate them into the technical report as desired.
- Have an outsider (i.e., someone who has not read any aspect of the report) conduct an editorial review to catch tone inconsistencies and grammar.

Academic Papers

For a subset of the findings from the Futurecasting Workshop, an academic paper can be written, especially if one of the gates is the requirement for more academic research on a specific topic. Publishing a paper or short article in an academic journal within that discipline could be very helpful for disseminating findings. Also, if you are a student or a new analyst, this could be the start of your publication history within your curriculum vitae (CV).

Writing for academic publications might be a slightly different style than you are used to.

Therefore, it can be helpful to find a writing mentor to help guide you through the process.

Ready to build an Academic Paper for The Project?

Take Action: Build an Academic Paper

- Write down novel ideas and findings that might work as the foundation for an academic paper.
- For each idea, brainstorm different participants and SMEs who might be good collaborators for the academic paper.
- Locate a writing mentor to help with the process, especially if academic writing is new to you.
- For each potential idea that you listed in #2, brainstorm a list of possible publications that might be appropriate for the paper.

Record the following information on these publication venues:

- Publication Name,
- Website,
- Submission Deadlines,
- Submission Criteria (including length, format, etc.)

Select one of the ideas and create an outline (that confirms with style guidance from the publication).

Briefings, BLUFs, Executive Summaries

A Bottom Line Up Front (BLUF), a summary, or a briefing offers a window into the high level of the analyst(s)' findings. The analyst(s) should always plan to create a briefing that summarizes the effort. Generally, the analyst should plan to offer this briefing to any person or organization that has been involved in the project. Aside from being a professional courtesy for participating, it is also an excellent way to socialize the findings and increase the social network for the analyst(s).

White Paper

Producing a white paper from a Futurecasting workshop is a valuable output for encapsulating complex future findings. A white paper can offer a detailed and authoritative format for presenting findings, making it useful for communicating nuanced opportunities to decision-makers. The structured nature of a white paper allows for clear articulation of methodologies, assumptions, and strategies, ensuring that the information is not only credible but also accessible for future reference and discussion. Additionally, the act of writing it encourages a deeper synthesis of the workshop's diverse ideas, solidifying the understanding and potential application of the Futurecasting outcomes.

8.7.1 Alternative Outputs

Events

Interactive events and experiences, like festivals and gatherings, provide immersive environments for engaging with future findings. These experiences allow participants to explore Futurecasting outcomes in a more hands-on manner, fostering deeper engagement and understanding.

Graphic Novellas/Comic Books

Graphic novels, novellas, and comic books are stories presented in a visual comic-strip format and published as a book. A graphic novella is a shorter version of a graphic novel. Those who do not have the time to read the entire technical report, nor a lengthy academic paper, will likely have the time to read a ten-page graphic novella. Typically, this will be enough to whet their interest in starting conversations about the topic within their community or organization or reading the other papers. Graphic novellas can be highly impactful because they can show the reader what a possible or potential future could look like. Because it is a visual medium, the writer and creators of the novellas can pack in a lot of information that was developed in the workshop and written about in the technical report.

Podcasts/Interviews/Presentations

An interview, podcast or a presentation might be an effective way to provide an audience with a high-level overview of the Futurecasting findings (similar to a written executive summary)—just in audio form. Recording a podcast or giving an interview prepares the core team to discuss the findings and to socialize them.

Physical Prototypes

Futurecasting workshops raw data can be used to build physical prototypes as an outcome in collaboration with worldbuilders. This approach enables participants to transform abstract future concepts into concrete models, offering a hands-on understanding of potential future technologies or solutions. Physical prototypes serve as a bridge between theoretical futures and their practical implications, allowing for a more comprehensive evaluation and exploration of ideas. They provide a visual and tactile representation of future concepts, which can be invaluable for communicating complex ideas. Building physical prototypes as an output of Futurecasting workshops plays a role in visualizing, testing, and refining future findings in a more immediate and relatable way.

Sci-Fi Prototypes

Science Fiction Prototype, is a narrative storytelling technique used to present possible futures and technology development. It involves crafting science fiction stories based on current scientific research and emerging technologies. The purpose of SFP is to explore potential future technologies, their applications, and the societal impacts they might have. By imagining possible futures through the lens of science fiction, researchers and innovators can better understand the implications of their work and guide the development of technology in a more informed, responsible, and ethical direction.

Scripts

Futurecasting workshops effects-based model (EBM) and data output provides a robust foundation for scriptwriting, particularly for world builders in fields such as game design, filmmaking, and speculative fiction. EBM's systematic approach to understanding the effects lends itself to the creation of rich, complex worlds and narratives. EBM helps in constructing detailed and plausible futures. These futures are grounded and offer a wealth of material for world builders to craft compelling narratives and settings. The data ensures that these fictional worlds are not only imaginative but also resonate. The transition from EBM outcomes to scripts involves harnessing the details to create narratives that are coherent, engaging, and reflective of the possible and potential futures.

An Interview

Madeline Ashby Consulting Futurist and Science Fiction Writer

My name is Madeline Ashby and I am a consulting futurist and science fiction writer and novelist, and I'm like a foresight Mary Poppins. I leave when the wind changes, I come in, I offer some perspective that might seem a little odd, um, but I try to leave things better than I found them, and I try to listen to people with my whole self. I try to understand that who we are at work is also who we can be at home. It's also who we were as children. It's also who we will be in the future. Um, I started out in a program at the, at ed. u, the Ontario College of Art Design University. I'm a graduate of the Strategic Foresight and Innovation Program there, and I was part of the initial cohort of that, uh, program. And what I wish we had taught students then. What I try to teach students now when I talk about strategic foresight, futuring, this kind of work, is that sitting with future possibilities is an embodied process. We do it with our whole body. It's not just a brain in a jar. It's not just an easily replicated algorithm, even. It is influenced by cortisol, stress, adrenaline, and hunger, whether or not you're a smoker. All kinds of things. A lot of foresight training will give you terminology. Very few will tell you to give your participants a meal break. The meal break is in fact, more critical. And learning how to read the room is more important. There's a vast amount of emotional intelligence that is required by this job. This job requires a vast amount of emotional labor. It is very draining. Um, it's also very reviving. You will leave workshops feeling more optimistic than you could, than you are right now.

I think that knowing that it was easy for me as a writer to understand that. But over the years, what I think I've learned also is that less is more. At the beginning, I really wanted to wow everybody by having no, by knowing all the words, knowing all the facts, knowing all the terminology that is good. Everyone should. Now, what I understand though, and it's a thing I learned from arts practice, from creative writing practice, from being a novelist, what I learned was the simpler you make things for people, the more generous you are being.

The act of creating, facilitating, leading, curating an experience for them in a workshop, in a written text, uh, in an installation certainly, or any kind of process, the more you take care of for them, the more generous you are being in holding back, we are actually leaving people, participants, you know, audiences, readers, whomever, space to fill with their own ideas, their own hopes, anxieties, fears. You have to leave them that room so that everything they've been holding back can actually surface. Because what will surprise you if you're just starting out, and which I think continues to surprise long-term practitioners, is that they hear truths or stories that participants have held back or sat on for far too long. And I wish that I had known at the beginning what, what a very sacred trust that would be, really, because you hear the things that other people are very worried to share.

And yet, those worries are often hugely important to their personal future, the future of their organization, how the team members of their department relate to one another. Those things are often the undergirding tension in why a person, a team, an organization, a company, whatever, can't move forward. This sounds like a, a big advertisement for therapy, because I often call strategic foresight corporate therapy. But it is about holding space for the dreams and fears, the hopes and anxieties, the optimism and the pessimism, all of it, of this group or the whoever it is that you're dealing with. And if they are willing to let you drill down, you'll find that it's often simpler than you think. That isn't to say that complexity doesn't exist. I don't believe in simplicity at the core of things. I think it depends what we go to college to learn, but I also think that when we find opportunities to relate to one another as human beings first, it becomes a lot easier to ask people to make a change.

Conclusion

> If you fall in love with the imagination, you understand that it is a free spirit. It will go anywhere, and it can do anything
>
> —Alice Walker

At the lab, we have a common saying: "the Process is the Process." We say this on repeat because it is our belief that the full value and understanding of Futurecasting emerges most effectively through repeat engagement with the methodology. Initially, newcomers to Futurecasting may find the process, particularly in the early stages of the effects-based model (EBM) round, somewhat cumbersome and disjointed. This sensation is not unusual, as the intricate steps of Futurecasting require a certain level of acclimatization.

However, as participants engage in multiple rounds, there is a noticeable shift in their approach and comprehension. The method, which might have once seemed perplexing, begins to unfold with greater clarity for the participants. Small teams and individual participants start to grasp the method, learning how to allocate their time and engage their imagination. They discover which aspects require careful deliberation and which can be navigated more swiftly.

This understanding and new skill of the nuanced design of the Futurecasting framework offers a structured way that each round enhances the participants' ability to foresee futures and deepens their appreciation of the process itself. In essence, by saying "the Process is the Process," we acknowledge that the journey through Futurecasting is as crucial as the outcomes it generates, providing invaluable insights and fostering a more profound grasp of the method with each iteration. This is also the reason we invite repeat "Futurecasters" to workshops as part of their role is ensuring newcomers to trust the process.

B. D. Johnson and C. Coon, *Futurecasting*, Synthesis Lectures on Threatcasting,
https://doi.org/10.1007/978-3-032-21993-0

With Great power comes Great Responsibility.

—Stan Lee

Now you have the ability to design and shape the future. The future is not a fixed point on the horizon toward which we are running, helpless to take action. We have the ability to shape the future. Futurecasting can help. But it is not the key drive of changing the future … you are.

Futurecasting is a human-centered method that draws from human knowledge and participation and needs human analysts to make sense of the data, stories and output. It means, in fact, that the future depends on you. The future is shaped and made better when passionate people get together and envision solutions, experience, and positive futures.

This now is your task.

That is our challenge to you.

You now have the ability to envision and shape the future. How will you use your passion and energy to make the future better? How will you reach out, inform others, and enlist others to help you? The future is a team sport. It is built one person at a time. That first person is you.

Useful Links

Multi-Sided Dice link Dice Shop: https://www.thediceshoponline.com/specialist-dice/multi-sided-dice-sets.html

Applied Future Lab: https://www.appliedfutureslab.com/

Threatcasting Community: Threatcasting.ai

Threatcasting Lab at ASU: https://threatcasting.asu.edu/

Glossary

AAR After Action Report An after action report (AAR) is a structured review or analysis conducted after the completion of a project, task, event, or situation. Its primary purpose is to evaluate what happened, why it happened, and how it can be done better by the people involved and those responsible for the project or event. AARs are commonly used in the military, emergency services, and business sectors for continuous improvement. They typically involve a detailed breakdown of events, assessment of objectives achieved, identification of successes and failures, and formulation of lessons learned and recommendations for future action.

BLUF Bottom-Line Up-Front Bottom line up front (BLUF) is a communication technique where the essential information or conclusion is presented at the beginning of the text, report, or briefing, followed by the supporting details. This approach is favored for its efficiency and clarity, ensuring that the main point is immediately understood by the reader or listener, even if they do not have the time or need to engage with the entire content. BLUF is widely used in military, business, and other contexts where clear, concise communication is critical.

CV Curriculum Vitae A curriculum vitae (CV) is a comprehensive document that outlines an individual's academic and professional history. Unlike a resume, which is typically brief and focused on specific skills and experiences relevant to a particular job, a CV is more detailed and encompasses one's entire career. It includes education, work experience, publications, awards, scholarships, grants, research projects, presentations, and other achievements. CVs are commonly used in academic, scientific, research, and medical professions, and they can be several pages long, providing a thorough overview of the individual's professional life.

EBM Effects-Based Model Effects-based model (EBM) refers to a strategic planning and analysis approach focusing on the outcomes or effects of actions, rather than just the actions themselves. In this model, the primary concern is understanding how specific actions will influence the behavior or status of systems or actors. EBM is often used in military strategy, policy-making, and business planning, where it's crucial to anticipate

B. D. Johnson and C. Coon, *Futurecasting*, Synthesis Lectures on Threatcasting,
https://doi.org/10.1007/978-3-032-21993-0

the broader impacts of decisions on the environment, the target audience, or the market. This approach helps in creating strategies that are more likely to achieve desired end-state objectives by considering the cascading effects of decisions.

EBO Effects-Based Operations Effects-based operations (EBO) is a military concept focused on achieving strategic goals by carefully planning and executing operations that consider the desired effects on an adversary or the environment. Instead of simply targeting enemy forces or resources, EBO involves understanding the enemy's systems and networks to apply actions that produce specific psychological, physical, functional, and systemic effects. This approach aims to disrupt or manipulate the enemy's decision-making process and capabilities, thereby achieving broader strategic objectives with potentially fewer resources and less conflict. EBO requires a deep understanding of the operational environment, the enemy, and the interdependencies within the system.

HCI Human Computer Interface Human–computer interface (HCL), also known as Human–Computer Interaction, is the study and practice of designing, implementing, and evaluating interactive computer systems from a user-centered perspective. It focuses on how people interact with computers and how to make this interaction more effective, efficient, and satisfying. HCI encompasses a wide range of topics, including user interface design, usability, ergonomics, and accessibility. It integrates concepts from computer science, psychology, design, and several other fields to create interfaces that are intuitive and responsive to users' needs and preferences.

IT Information Technology Information technology (IT) is the use of computers, networking, and other physical devices, infrastructure, and processes to create, process, store, secure, and exchange all forms of electronic data. Typically, IT is used within the context of business operations as opposed to personal or entertainment technologies. The IT field encompasses a wide range of areas including software development, computer systems architecture, and information systems. It plays a critical role in data management, telecommunications, and provides support for decision-making processes within organizations.

PII Personally Identifiable Information Personally identifiable information, refers to any data that can be used to identify a specific individual. This can include direct identifiers, like names and social security numbers, that can uniquely identify a person, as well as indirect identifiers, like race, birthdate, or geographic indicators, that can be combined with other information to identify a person. Protecting PII is crucial in maintaining individual privacy, and it is a key focus in data security practices to prevent identity theft and unauthorized access to personal information.

RSW Research Synthesis Workbook The Research Synthesis Workbook (RSW) is a tool or document used in academic and professional research to organize, analyze, and synthesize findings from various studies or sources. It's designed to help researchers systematically review and integrate data and insights from a range of research materials. The workbook format allows for a structured approach to identifying patterns, discrepancies, and key themes across different studies, facilitating a deeper understanding of the research topic. RSWs are particularly useful in fields where comprehensive reviews of existing literature are essential for advancing knowledge or informing practice and policy.

SFP Science Fiction Prototype Science fiction prototype, is a narrative storytelling technique used to present possible futures and technology development. It involves crafting science fiction stories based on current scientific research and emerging technologies. The purpose of SFP is to explore potential future technologies, their applications, and the societal impacts they might have. By imagining possible futures through the lens of science fiction, researchers and innovators can better understand the implications of their work and guide the development of technology in a more informed, responsible, and ethical direction.

SME Subject Matter Expert Subject matter expert, is an individual with a deep understanding and specialized knowledge in a particular area or topic. They are considered authorities in their field, often due to a combination of education, experience, and skills. SMEs are critical in providing insights, detailed information, and expert opinions in their area of expertise. They are often consulted in projects, research, and decision-making processes to ensure accuracy, credibility, and depth of understanding. Their roles can vary across different industries, from technical fields like engineering or IT to academic research and business consulting.

TAD Time-Phased, Alternative-action Definition TAD, or time-phased, alternative-action definition, is a strategic planning and analysis concept. It involves breaking down a project or initiative into distinct phases over time and defining alternative courses of action for each phase. This approach allows for flexibility and adaptability in planning, as it considers different scenarios and potential changes in circumstances. By outlining alternative actions at different time phases, decision-makers can better prepare for uncertainties and make informed choices that are responsive to evolving situations. TAD is particularly useful in complex projects or environments where predicting future conditions with certainty is challenging.

UX User Experience UX, or user experience, refers to the overall experience of a person using a product, system, or service, especially in terms of how easy or pleasing it is to use. It encompasses all aspects of the end-user's interaction with the company, its services, and its products. The primary goal of UX design is to create products that provide meaningful and relevant experiences to users. This involves the design of the entire process of acquiring and integrating the product, including aspects of branding, design, usability, and function. UX is a critical aspect in the development of websites, applications, software, and virtually any product that requires user interaction, focusing on optimizing satisfaction and usability.

Bibliography

Army Cyber Institute (2018) Dark hammer: a retrospective of science fiction prototyping. Retrieved from https://Futurecasting.asu.edu/Dark_Hammer_Retrospective

Atkinson RC, Shiffrin RM (1968) Human memory: a proposed system and its control processes. In: Psychology of learning and motivation, vol 2. Academic Press, pp 89–195. https://doi.org/10.1016/S0079-7421(08)60422-3

Basso D, McCoy N (1996) Study tools: a comprehensive curriculum guide for teaching study skills to students with special needs. Twins Publication, Columbia, SC. https://eric.ed.gov/?id=ED412700

Bennett M, Johnson BD (2016) Dark future precedents: science fiction, futurism and law. In: Intelligent environments (Workshops), pp 506–513

Bradberry R (1950) August 2026: there will come soft rains. In: Martian chronicles. Simon and Schuster Paperbacks, New York, NY

Bradfield RM (2008) Cognitive barriers in the scenario development process. Adv Develop Human Resour 10(2):198–215. https://doi.org/10.1177/1523422307313320

Brown J (2021) Thinking like a futurist: investigating the theories and processes of futurecasting post-analysis. Doctoral dissertation, Arizona State University

Cabrera D, Colosi L, Lobdell C (2008) Systems thinking. Eval Program Plann 31(3):299–310. https://doi.org/10.1016/j.evalprogplan.2007.12.001

Cole A, Singer PW (2015) Ghost fleet: a novel of the next world war. Eamon Dolan/Houghton Mifflin Harcourt

Cole A, Singer PW (2020) Burn-in: a novel of the real robotic revolution. Mariner Boooks

Cowan N, Elliott EM, Saults JS, Morey CC, Mattox S, Hismjatullina A, Conway AR (2005) On the capacity of attention: its estimation and its role in working memory and cognitive aptitudes. Cogn Psychol 51(1):42–100. https://doi.org/10.1016/j.cogpsych.2004.12.001

Defense Acquisition University (n.d.) DOTmLPF-P analysis. Retrieved from https://www.dau.edu/acquipedia/pages/ArticleContent.aspx?itemid=457

Desjardins J (2017 Sept 25) Every single cognitive bias in one infographic. Retrieved from https://www.visualcapitalist.com/every-single-cognitive-bias/

Dweck CS, Leggett EL (1988) A social-cognitive approach to motivation and personality. Psychol Rev 95(2):256. https://doi.org/10.1037/0033-295X.95.2.256

Eccles J (1983) Expectancies, values, and academic behaviors. In: Spence JT (eds) Achievement and achievement motives: psychological and social approaches. W. H. Freeman, New York, NY

Ellis S (2008 Apr 10) Frameworks, methodologies and processes. Retrieved from http://vsellis.com/frameworks-methodologies-and-processes/

Feynman R (2006) The Feynman lectures on physics [UNABRIDGED]. Audio CD. Basic Books

B. D. Johnson and C. Coon, *Futurecasting*, Synthesis Lectures on Threatcasting,
https://doi.org/10.1007/978-3-032-21993-0

Gabriel D (2011 May 13) Methods and methodology. Retrieved from https://deborahgabriel.com/2011/05/13/methods-and-methodology/#:~:text=Method%20is%20simply%20a%20research,using%20a%20particular%20research%20method

Green K, Vergragt P (2002) Towards sustainable households: a methodology for developing sustainable technological and social innovations. Futures 34(5):381–400. https://doi.org/10.1016/S0016-3287(01)00066-0

Hojer M, Gullberg A, Pettersson R (2011) Backcasting images of the future city—time and space for sustainable development in Stockholm. Technol Forecast Soc Change 78(5):819–834. https://doi.org/10.1016/j.techfore.2011.01.009

Inayatullah S (2008) Six pillars: futures thinking for transforming. Foresight 10(1):4–21. https://doi.org/10.1108/14636680810855991

Incrementalism (2021 June 10) In Britannica. https://www.britannica.com/topic/incrementalism

Joint Chiefs of Staff (2021) Department of defense dictionary of military and associated terms. United States Government. Retrieved from https://www.jcs.mil/Portals/36/Documents/Doctrine/pubs/dictionary.pdf

Johnson BD, Vanatta N (2017) What the heck is Futurecasting? Future tense. Retrieved from https://Futurecasting.asu.edu/sites/default/files/2019-11/what-the-heck-is-Futurecasting.pdf

Johnson BD, Vanatta N, Draudt A, West JR (2017) The new dogs of war: the future of weaponized artificial intelligence. Retrieved from https://apps.dtic.mil/sti/citations/AD1040008

Johnson BD (2018) Hero. Army Cyber Institute, west point. Retrieved from https://Futurecasting.asu.edu/graphic-novella/hero-2028

Kolb DA (2014) Experiential learning: experience as the source of learning and development, 2nd edn. FT press

Lindblom CE (1959) The science of "muddling" through. Public Administrat Rev 19(2):79–88. Retrieved from https://www.jstor.org/stable/973677. https://doi.org/10.2307/973677

Lovin AB (1976) Energy strategy: the road not taken. Foreign Affairs 55:65–96. https://doi.org/10.2307/20039628

Markowitz E (2020, Feb 6) How to write an executive summary. Retrieved from https://www.inc.com/guides/2010/09/how-to-write-an-executive-summary.html

Maslow AH (1943) A theory of human motivation. Psychol Rev 50(4):370. https://doi.org/10.1037/h0054346

Meissner P, Wulf T (2013) Cognitive benefits of scenario planning: its impact on biases and decision quality. Technol Forecast Soc Change 80(4):801–814. https://doi.org/10.1016/j.techfore.2012.09.011

MindTools (n.d.) Paraphrasing and summarizing. Retrieved from https://www.mindtools.com/pages/article/paraphrasing-summarizing.htm

Nenniger P (1992) Task motivation: an interaction between the cognitive and content-oriented dimensions of learning. In: Renninger KA, Hidi S, Krapp A (eds) The role of interest in learning and development. Psychology Press, New York, NY, pp 121–150

Perception Institute (n.d.) Implicit bias Retrieved from https://perception.org/research/implicit-bias/

Project Implicit (n.d.) Project implicit. Retrieved from https://www.projectimplicit.net/

Renninger KA, Hidi S, Krapp A, Renninger A (eds) (2014) The role of interest in learning and development. Psychology Press. https://doi.org/10.4324/9781315807430

Ringland G, Schwartz PP (1998) Scenario planning: managing for the future. Wiley

Robinson JB (1982) Energy backcasting: a proposed method of policy analysis. Energy Policy 10(4):337–344. https://doi.org/10.1016/0301-4215(82)90048-9

Robinson JB (1990) Futures under glass: a recipe for people who hate to predict. Futures 22(8):820–842

Shonka M, Kosch D (2002) Beyond selling value: a proven process to avoid the vendor trap. Dearborn Trade Publishing. https://www.amazon.com/Beyond-Selling-Value-Proven-Process/dp/0793154707

Staples WG (ed) (2007) Encyclopedia of privacy: AM, vol 1. Greenwood Publishing Group, p 93

Tuominen A, Tapio P, Varho V, Jarvi T, Banister D (2014) Pluralistic backcasting: integrating multiple visions with policy packages for transport climate policy. Futures 60:41–58. https://doi.org/10.1016/j.futures.2014.04.014

Tversky A, Kahneman D (1982) Judgement under uncertainty: heuristics and biases. In: Kahneman D, Slovic P, Tversky A (eds) Judgement under uncertainty: heuristics and biases. Cambridge University Press, New York, NY, pp 3–10. https://doi.org/10.1017/CBO9780511809477.002

van de Kerkhof M, Cuppen E, Hisschemoller M (2009) The repertory grid to unfold conflicting positions: the case of a stakeholder dialogue on prospects for hydrogen. Technol Forecast Soc Change 76(3):422–432. https://doi.org/10.1016/j.techfore.2008.07.004

The manufacturer's authorised representative in the EU is Springer Nature Customer Service Centre GmbH, Europaplatz 3, 69115 Heidelberg, Germany. If you have any concerns regarding our products, please contact ProductSafety@springernature.com

Printed and bound by CPI Group (UK) Ltd, Croydon, CR0 4YY
16/09/2026
02217780-0016